International Science and Technology Conference for Youth "Advanced Materials for Engineering and Functional Purposes" (AMEFP 2020)

IOP Conference Series: Materials Science and Engineering
Volume 1093

Tomsk, Russia
21 - 25 September 2020

ISBN: 978-1-7138-3350-5
ISSN: 1757-8981

Printed from e-media with permission by:

Curran Associates, Inc.
57 Morehouse Lane
Red Hook, NY 12571

CURRAN ASSOCIATES INC. proceedings .com

Some format issues inherent in the e-media version may also appear in this print version.

Additional copies of this publication are available from:

Curran Associates, Inc.
57 Morehouse Lane
Red Hook, NY 12571 USA
Phone: 845-758-0400
Fax: 845-758-2633
Email: curran@proceedings.com
Web: www.proceedings.com

TABLE OF CONTENTS

IOP Conf. Series: Materials Science and Engineering 1093 (2021) 011001 doi:10.1088/1757-899X/1093/1/011001

International Science and Technology Conference for Youth "Advanced Materials for Engineering and Functional Purposes" (AMEFP-2020) 21 – 25 September 2020, Tomsk, Russia

Tomsk Polytechnic University (TPU)

Since 2003 Faculty of Mechanical Engineering (now School of Advanced Manufacturing Technologies) of Tomsk Polytechnic University periodically organizes International Science and Technology Conference "Advanced Materials for Engineering and Functional Purposes". Today TPU, relying on its unique experience and scientific developments, continues to perform the strategic objective of training personnel for industry facilities not only in Russia, but also abroad.

The International Science and Technology Conference for Youth "Advanced Materials for Engineering and Functional Purposes" (AMEFP-2020) was successfully held from September 21th to September 25th, 2020. The Conference gathered young scientists, students, researchers, and invited industrial specialists working in manufacturing, engineering, physical-chemical and information technologies, as well as in automation systems, materials science and medicine in order to facilitate aggregation and sharing interests and results for a better collaboration in advanced manufacturing technologies and development.

The major topics included:

- Problems of structural materials reliability
- Functional materials
- Surface modification

We want to thank all members of the Scientific Committee, Organizing Committee, Tomsk Polytechnic University, Institute of Strength Physics and Materials Science of Siberian Branch of Russian Academy of Sciences and everyone who contributed to the organization of this event, for their contribution towards the conference and for their contributions to this proceeding.

The Organizing Committee of AMEFP-2020

Scientific Committee

Prof Sergei Panin – Chairman, Head of Laboratory of Polymer Composite Materials, Institute of Strength Physics and Materials Science, Siberian Branch, Russian Academy of Sciences Tomsk, Russia
Prof Vasiliy Klimenov – Co-Chairman, Head of the Chair of Materials Science, Tomsk Polytechnic University, Tomsk, Russia
Dr Alexey Yakovlev, Rector of Kuzbass State Technical University, Kemerovo, Russia
Prof Sergei Kul'kov, Head of Laboratory of Physics of Nanostructured Ceramic Materials, Institute of Strength Physics and Materials Science, Siberian Branch, Russian Academy of Sciences, Tomsk, Russia
Prof Anatoli Bataev, Rector of Novosibirsk State Technical University, Novosibirsk, Russia
Prof Sunder Ramasubbu, Research Director of BISS (P) Ltd · R&D, Bangalore, India
Prof Shan Dong Tu, East China University of Science and Technology, Shanghai, China
Prof Filippo Berto, Norwegian University of Science and Technology, Trondheim, Norway
Prof Laszlo Gomza, Head of department of Non-metallic Materials Technology, University of Miskolc, Hungary

Organizing Committee

Prof Svetlana Buyakova – Chairman, professor of Tomsk Polytechnic University and Tomsk State University, Tomsk, Russia
Dr Olga Vaulina – Co-Chairman, associate professor, Tomsk Polytechnic University, Tomsk, Russia
Inessa Vasileva, senior lecturer, Tomsk Polytechnic University, Tomsk, Russia
Dr Mikhail Burkov, associate professor, Tomsk Polytechnic University, Tomsk, Russia
Dr Boris Zenin, associate professor, Tomsk Polytechnic University, Tomsk, Russia
Dr Elena Darenskaya, associate professor, Tomsk Polytechnic University, Tomsk, Russia
Dr Alexey Kondratyuk, associate professor, Tomsk Polytechnic University, Tomsk, Russia
Dr Elena Dedova, associate professor, Tomsk Polytechnic University, Tomsk, Russia

Editors

Prof Evgeni Kolubaev, Director of Institute of Strength Physics and Materials Science, Siberian Branch, Russian Academy of Sciences, Tomsk, Russia
Prof Sergei Panin, Head of Laboratory of Polymer Composite Materials, Institute of Strength Physics and Materials Science, Siberian Branch, Russian Academy of Sciences, Tomsk, Russia
Dr Ekaterina Kulinich, academic secretary of School of Nuclear Science & Engineering, Tomsk Polytechnic University, Tomsk, Russia

IOP Conf. Series: Materials Science and Engineering 1093 (2021) 011002 doi:10.1088/1757-899X/1093/1/011002

Peer review declaration

All papers published in this volume of *IOP Conference Series: Materials Science and Engineering* have been peer reviewed through processes administered by the Editors. Reviews were conducted by expert referees to the professional and scientific standards expected of a proceedings journal published by IOP Publishing.

- **Type of peer review:** Single-blind

- **Conference submission management system:** Papers were submitted via web-site of the Conference and e-mail kulinich@tpu.ru

- **Number of submissions received:** 68

- **Number of submissions sent for review:** 63

- **Number of submissions accepted:** 34

- **Acceptance Rate (Number of Submissions Accepted / Number of Submissions Received X 100):** 51

- **Average number of reviews per paper:** 2

- **Total number of reviewers involved:** 7

  Any additional info on review process:
  For the assessment of the papers editors used technical, quality and presentation criteria. Conference organizers assigned at least two reviewers for each paper.
  Reviewers' identities were hidden from authors (single-anonymous).
  The reviewers provided an overall evaluation (strong accept – accept – weak accept – borderline paper – weak reject – reject – strong reject) and decision to publish the paper as is, after minor/major revision or recommendation to reject the paper in this format.
  Reviews also included comments for authors. Authors had the opportunity to edit the paper according to the comments within 1 month. If reviewers agreed with the changes, the article was considered as accepted.

- **Contact person for queries:** Dr. Ekaterina Kulinich, kulinich@tpu.ru

Numerical study of the influence of relative parameters of the metamaterial structure on its mechanical behavior

L R Akhmetshin[1,2]

[1]Institute of Strength Physics and Materials Science of Siberian Branch of Russian Academy of Sciences, 2/4 Akademicheskii pr., Tomsk, 634055, Russia
[2]Tomsk State University, 36 Lenin ave., Tomsk, 634050, Russia

E-mail: akhmetshin.lr@gmail.com

Abstract. In this paper, attention is focused on the influence of changes in the variable parameters of the structure of mechanical tetrachiral metamaterial on its linear elastic behavior, in particular, on its twist. The parameters characterizing the structure of the metamaterial were chosen in a relative form with respect to the unit cell size and changed independently of each other in the investigation. The results of the mechanical behavior of the tetrachiral metamaterial in the event of changes in the structural parameters were obtained. The dependencies of the rotation angle when the relative parameters change were established and analyzed. The parameters of the chiral structure, the most affecting the unusual behavior of mechanical metamaterial—a twist under uniaxial loading, were determined.

1. Introduction

The mechanical metamaterials are a kind of metamaterials rapidly developing over the past ten years. Metamaterials are materials whose properties are determined by their artificially created structure but not by their chemical composition.

The interest in the development of mechanical metamaterials is partly due to advances in additive manufacturing technologies, which allowed the production of materials with an arbitrarily complex micro/nanostructure. The desire to create mechanical metamaterials is associated with the creation of structures with a certain set of physical and mechanical properties.

In comparison to the natural materials which have congenital mechanical properties at different scale levels, metamaterials are characterized by construction-specific properties at high length scales and, if we look at their behavior at the scale of the unit cell (micro and nanoscales), it corresponds to the behavior of the base material. However, in general, the average macroscale mechanical behavior of metamaterials should have some unusual effects [1].

Specific classes of mechanical metamaterials were known already a few decades ago. In recent years, metamaterials with a negative Poisson's ratio (auxetics) have been gaining popularity [2]. They are interesting because exhibiting unusual behavior—expand transversely in the tension and compression. Specific examples of such materials have been observed, created, tested, and reported elsewhere [3, 4].

One type of auxetic metamaterials is the materials with chirality of structure. Chiral metamaterials have several known advantages, such as an additional degree of freedom under the uniaxial loading—twist (rotation) [5], a large volume of pores, which depend on the parameters of the structure of the chiral metamaterial [6].

In this paper, we will pay special attention to the influence of changes in the variables of relative parameters of the structure of mechanical tetrachiral metamaterial on their elastic behavior, in particular, on the twist.

2. Mathematical stage

2.1. Geometry model

Since the emphasis in this paper is placed on the structure of the metamaterial, it is worth saying that the tetrachiral structure is a structure consisting of a ring and 4 ribs tangentially connected to the ring (figure 1a). The resulting structure is obtained in the plane, i.e. is a two-dimensional structure. To obtain the three-dimensional metamaterial it is necessary to build an elementary cell composed of the two-dimensional structures (figure 1b). Then a sample of metamaterial was made up of the elementary cells (figure 1c) by connecting the cells in three orthogonal axes according to the following numbers 3–3–9 in the corresponding axes X–Y–Z. The total number of cells in the sample equals $n = 91$.

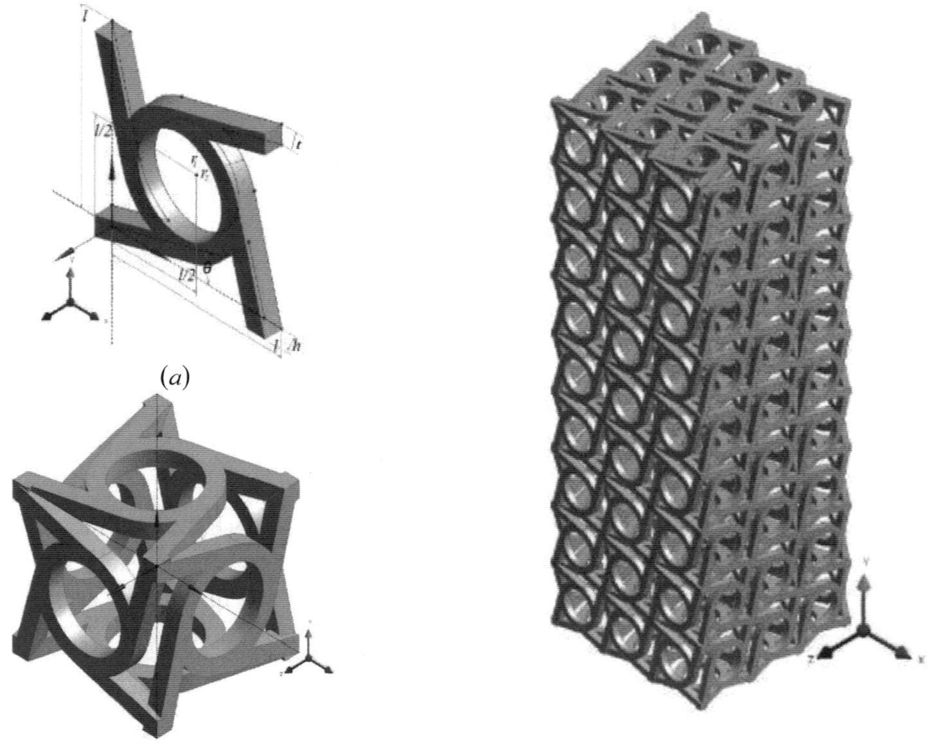

(*a*) (*b*) (*c*)

Figure 1. The structure of the mechanical tetrachiral metamaterial.

The parameters of the chiral structure were changed independently of each other. For example, when a specific parameter was changed, the others took the original values. The initial values of the parameters of the structure of mechanical tetrachiral metamaterial are presented in table 1, and their notions are shown in figure 1a.

Table 1. Initial parameters of the chiral metamaterial structure.

l (mm)	t (mm)	h (mm)	r_1 (mm)	r_2 (mm)	θ (°)
50	5	5	12.5	17.5	25

Here *l/2* is the unit cell length, *t* is the rib width, *h* is the rib thickness, r_1 is the inner radius of the ring, r_2 is the outer radius of the ring, θ is the angle of the inclination of the rib to the horizontal axis.

2.2. Mathematical model

The quasi-static problem of solid mechanics within the limits of the linear theory of elasticity was solved using the software package ANSYS. Hooke's law was chosen as the constitutive relation:

$$\sigma_{ij} = \lambda \cdot \delta_{ij} \cdot \varepsilon_{kk} + 2 \cdot \mu \cdot \varepsilon_{ij},$$

where $\varepsilon_{kk} = \varepsilon_{11} + \varepsilon_{22} + \varepsilon_{33}$, $\lambda = \dfrac{E \cdot \vartheta}{(1+\vartheta)(1-2\cdot\vartheta)}$, $\mu = \dfrac{E}{2\cdot(1+\vartheta)}$ are the Lame coefficients, δ_{ij} is the Kronecker symbol, σ_{ij} is the stress tensor, ε_{ij} is the strain tensor, E is the Young's modulus, ϑ is the Poisson's ratio. The following values of elastic moduli were used in this paper $E = 200$ GPa, $\vartheta = 0.3$.

2.3. Boundary conditions

To simulate the uniaxial loading of the sample, the corresponding boundary conditions were applied to the lower (rigid fixation) and upper (displacement constraint in the *Y*-axis and free motion in the *X* and *Z* axes) boundaries:

$$U_x^{fix} = U_y^{fix} = U_z^{fix} = 0$$
$$U_x^{disp} = U_x^{disp} = free$$
$$U_y^{disp} = -15 \approx -450 \cdot 3.33\%$$

The negative displacement given at the top of the metamaterial sample corresponds to compression. The absence of the minus sign would indicate that the metamaterial is stretched. From the last record of the boundary conditions, it can be understood that the deformation of metamaterials does not exceed 4%.

3. Results and discussion

The results of the influence of variable parameters of the cell structure of mechanical tetrachiral metamaterials were described in detail in [6]. Now, we will focus on the invariant relationships of the geometric model to make the structure dimensionless. Thus, we solve the problem of scalability of the metamaterial sample.

The cell length is taken as the basic value of the unit cell. This parameter will be in the denominator and the relative values look like r_1/l, t/l, h/l, θ/l. The angle θ characterizes the change of the outer ring radius (r_2).

By the form of the dependencies and the range of the angle α in figure 2a, it is clearly seen that the mechanical response we are interested in—a twist—is rather similar when we change the inner radius of the ring and the width of the ribs. The twist angle of the sample having the initial parameters of the structure is obviously the same. The corresponding results for the maximum change in each of the considered parameters result in almost identical twist of the sample.

If one compares the influence of the two parameters of the rib section, namely the width and the thickness, the great difference will be found in the plot in figure 2a. Changing in the rib width has almost no effect on the sample twist. It comes as a surprise because the corresponding changes in the structural parameters of the metamaterial have approximately identical impact on the volume of the base material. Probably, this difference is due to the effect of bending forces on the rib structure of the metamaterial. Rib width is an external parameter for the in-plane chiral structure and has much less effect on the bending forces than the rib thickness. As a consequence, the twisting of the metamaterial is scarcely affected by the changing in rib width. This effect can be attributed also to the fact that uniaxial loading of the sample affects the bending in that plane in which the change in thickness is significant but not in the width.

The angle of inclination of the rib to the *X*-axis is the parameter that most significantly affects the twist of the metamaterial sample, as it is responsible for the twisting of the ring (figure 2b). Decrease of the twist angle with increasing the relative parameter of the rib angle passes on the parabolic law, to the absolute zero value of the twist angle which corresponds to the value of $\theta/l = 0.9$ °/mm.

The further increase in the described parameter will lead to a further parabolic decrease of the twist angle into the region of negative values. That means that the mechanical tetrachiral metamaterial sample will twist in the opposite direction.

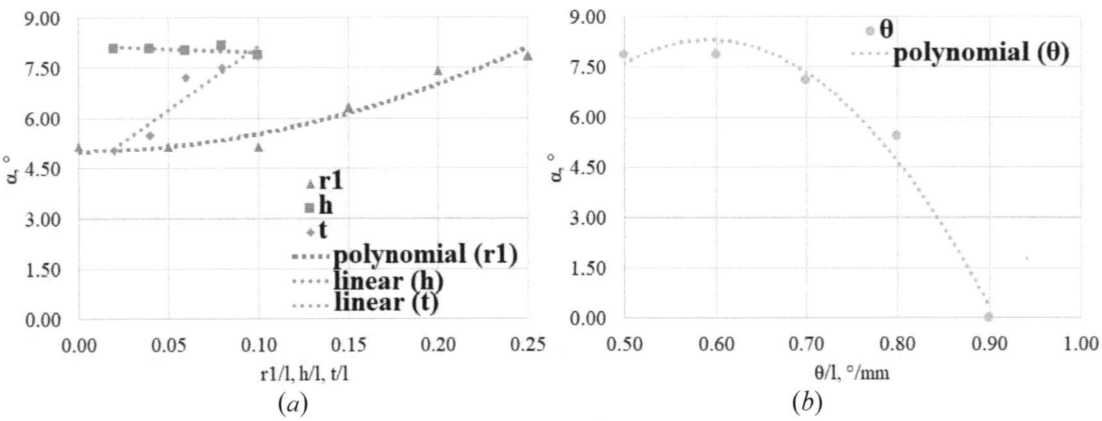

Figure 2. Twist angle of the metamaterial as a function of the variable relative parameters.

4. Conclusion

In this work, the mechanical behavior of the tetrachiral mechanical metamaterial sample was studied under uniaxial loading. The problem was treated as a quasi-static problem in the framework of solid mechanics. Numerical simulation of the metamaterial mechanical behavior was carried out within the framework of the linear theory of elasticity. The deformation of the metamaterial did not exceed 4%.

The parameters characterizing the structure of the metamaterial were presented in a relative form and varied independently. The dependencies of the twist angle on the changing of the relative parameters were obtained. The parameter of the chiral structure most affecting the unusual behavior of the mechanical metamaterial—a twist under uniaxial loading, is identified.

When increasing some parameters, the twist angle decreases. This can be associated with an increase in relative volume by a trend: "the larger the relative volume of metamaterial, the smaller the twist angle". This trend is not valid for all parameters.

Acknowledgments
The work was performed according to the Government research assignment for ISPMS SB RAS, project No. III.23.2.12.

References

[1] Zadpoor A A 2016 *Mater Horizons* **3**(5) 371-381
[2] Evans K E, Nkansah M A, Hutchinson I J and Rogers S C 1991 *Nature* **353**(6340) 124-125
[3] Goldstein R V, Gorodtsov V A, Lisovenko D S and Volkov M A 2014 *Phys. Mesomech.* **17**(2) 97-115
[4] Goldstein R V, Gorodtsov V A, Lisovenko D S and Volkov M A 2019 *Phys. Mesomech.* **22**(4) 261–268
[5] Frenzel T, Kadic M and Wegener M 2017 *Science* **358**(6366) 1072-1074
[6] Akhmetshin L R and Smolin I Yu 2020 *Nanosci. Technol: Int. J.* **11**(3) 265-273

Numerical study of hard-metal powder compaction

L R Akhmetshin[1,2], I Yu Smolin[1,2] and S P Buyakova[1,2]

[1]Institute of Strength Physics and Materials Science of Siberian Branch of Russian Academy of Sciences, 2/4 Akademicheskii pr., Tomsk, 634055, Russia
[2]Tomsk State University, 36 Lenin ave., Tomsk, 634050, Russia

E-mail: akhmetshin.lr@gmail.com

Abstract. Powder compacting is a widely used process for both metallurgy and pharmaceutical production. The mathematical simulation of the powder compaction process is a promising tool for its study. The compacting of WC/Co powder by a punch in a cylindrical die is simulated in this work. The use of the Drucker-Prager Cap constitutive model is made. Results on stress distribution and volume plastic deformation during compaction are obtained. Relative density distribution in the powder compact can also be estimated from volume plastic strain. The severe influence of frictional contact between powder and die wall on the results is noted.

1. Introduction

Powder compaction is an easy and cheap process that is widely used for the production of materials and industrial components [1]. In the manufacturing technologies of powder metallurgy (PM) products, cold and hot isostatic pressing and die compaction are widely used. One of the main problems in such technologies is that PM parts formed by die compaction have inhomogeneous density distributions due to the friction between the powder and the die wall. Inhomogeneity in density leads to non-uniform shrinkage or distortion during the powder compact sintering, which may lead to cracking of the obtained products [2].

PM is a fast-growing segment of the industry as, for example, automotive manufacturers look to reduce the overall weight of vehicles by replacing a range of ferrous components with PM products and thus increase car fuel efficiency.PM products are looked to as feasible substitutes in the place of both die-cast aluminum and iron materials for moving engine components [3].

Active users of powder compacting technology are pharmaceutical companies. The compressing of dry powder mixtures is one of the most significant operational processes in the pharmaceutical industry. With the use of this technology, pills are usually made. A critical step in the manufacture of tablets is the compaction of a loose powder/granule formulation between two punches within a die. Among other qualities, the resulting compact should have nearly uniform properties with sufficient mechanical strength to avoid breakage and attrition [4].

Concerning hard metal components, they are widely used in different industries as cutting tool elements. For example, cutting inserts that are used as the cutting element in a lathe, especially in a situation where space is limited, are produced by powder metallurgy technology [5].

However, the final microstructure and the properties of the compacts depend on not only the choice of the powder composition. The choice of the parameters of the compaction process influences them as well. The identification of these parameters usually needs extensive experimental works that are tedious, expensive, and time-consuming.

Therefore, the numerical simulation using the finite element method has been employed as an alternative tool to investigate powder compacting processes over the past decades. This method can predict the distribution of variables such as density, strain, and stress in the powder mixture before the actual component design and manufacturing. This gives a better understanding of the compaction process and the influence of various parameters. However, the successful simulation requires a thorough knowledge of a constitutive model or a yield criterion, friction between die and powder, and well-defined initial state and conditions of the powder [1]. Moreover, simulation of the process compaction can also help to understand the influence of tooling properties, lubrication, and compaction kinematics, such as speed compaction and compaction sequences, and provide guidance for the optimization of tooling design and the improvement of the powder mixture [6].

The aim of this work is to study the process of compacting the mixture of hard-metal powders, as well as the distribution of stresses and bulk plastic deformations in the resulting compact.

2. Mathematical model of compaction

The phenomenological models, such as critical state plasticity models like Cam-Clay and Drucker-Prager Cap (DPC), which were originally developed for geological materials in soil mechanics, turned out to be well suited for modeling powder compaction, especially in PM. Recently, the DPC plasticity model has been widely used for powder compacting analysis of different powders as it can describe powder compaction and hardening.

The DPC model is one of the continuum mechanical models, in which the powder is considered as a porous medium. The DPC model can represent the densification and hardening of the powder, as well as the inter-particle friction. This model can also reasonably represent both the shear failure and the plastic yielding of the powder and can be readily characterized via experiments on real powders. Therefore, the DPC model is used frequently to analyze the strain, relative-density changes, and stress distribution of compacts during the compacting process. Finally, the DPC model is characterized by parameters that are mainly related to the internal friction angle, elastic modulus, and plastic deformation, among other parameters, that change through the hardening law [7].

Although the DPC model is a subset of the extended Drucker-Prager material, the Cap model should be viewed as a separate constitutive model. It has been proposed to enable users to describe the plastic volume compression of porous bodies when pores are collapsed and the subsequent hardening of the compressible material. With the regular Drucker-Prager models, the yield strength would increase indefinitely if the hydrostatic compressive stresses kept increasing in magnitude. A compaction cap allows users to define a hydrostatic pressure at which compaction occurs—hence, instead of yielding on the shear envelope, yielding can also occur from compaction or expansion [8].

Here, the ANSYS LS-DYNA software package was used to model the powder compacting process. In this software package, the model of Drucker-Prager Cap material is already implemented under the name *MAT_025/*MAT_GEOLOGIC_CAP_MODEL.

2.1. Geometry model

The geometry of the punch, the die, and the powder was constructed using Ansys DM in a two-dimensional axisymmetric description. It is assumed that both the punch and the die are rigid cylindrical bodies made of steel, and the powder is an elastic-plastic solid described by the DPC model. The cylindrical die has the inner radius of L1 = 7.5 mm and the outer radius of L2 = 10.5 mm (see the notations in figure 1). The initial filling height of the powder is H2 = 46.87 mm. The other values of geometry schematic depicted in figure 1 were as follows, H1 = H3 = 3 mm. The mathematical model was solved using axially symmetric quadrangular finite elements. The number of elements in the model was equal to 2241.

2.2. Material model parameters

In our work, we focus on modeling the process of compacting hard-metal powders. The composition of hard-metal powder varies depending on its application. For a cemented carbide a normal

IOP Conf. Series: Materials Science and Engineering 1093 (2021) 012002 doi:10.1088/1757-899X/1093/1/012002

composition is about 90 % of WC and about 10 % of cobalt, though other materials could also be included.

The parameters presented in table 1 were used to describe the material model. They were determined based on the data of Reference [9] for a precompacted powder of 15 MPa. The author of [9] derived the parameters from experimental data on three-axis loading.

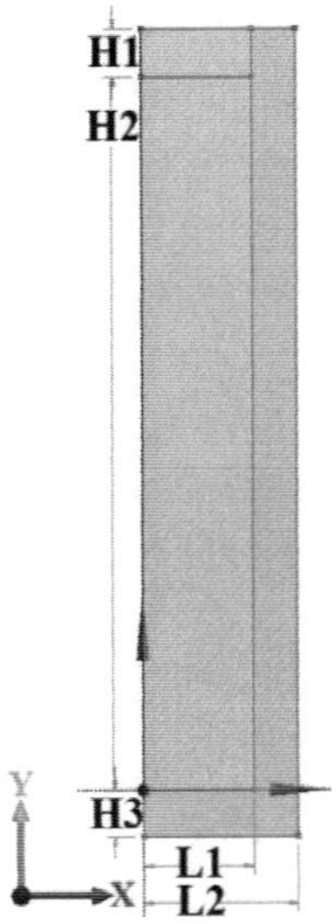

Figure 1. The geometry schematic.

Table 1. The Drucker-Prager Cap model parameters.

Parameter (Units)	Value
Initial density (kg/m^3)	6700
Initial bulk modulus (MPa)	701.1
Initial shear modulus (MPa)	593.2
Failure envelope parameter, α (MPa)	929.2
Failure envelope linear coefficient, θ	0.0
Failure envelope exponential coefficient, γ (MPa)	928.8
Failure envelope exponent, β (GPa^{-1})	0.051
Cap surface axis ratio, R	17.1
Hardening law exponent, D (MPa^{-1})	0.0074
Hardening law coefficient, W	0.3015

2.3. Boundary condition

The die is fixed in both the axial and the radial direction, while the punch is fixed only in the radial direction but free to move in the axial direction. The punch is loaded by smoothly increasing force to simulate the hydraulic press up to the pressure of 50 MPa. The contact between the die, the punch, and the powder is simulated by the algorithm of surface-to-surface contact in LS-DYNA and uses Coulomb friction with the value 0.16 taken from the literature [2, 9].

3. Results and discussion

In the phenomenological approach used, a pre-compacted powder mixture is considered as a continuous medium. The task of the study is to determine the distribution of stresses and plastic strains of this medium at different stages of the process, which allows assessing the quality of the resulting compact and the reasons for its deterioration.

In the model, the highest stresses are localized in the corner where all three elements of the model—the punch, die, and powder—contact each other, while the lowest stresses are in the opposite corner along the vertical axis. Stress gradients, as a rule, are the highest near the sidewall boundaries due to the resistance to powder movement resulting from friction with the wall.

The distributions of effective plastic strain during the process are presented in figure 2. It may be noted that starting the initial stage of compacting the plastic strains are distributed non-uniformly throughout the powder. As the time of the process increases, also increases the localization of strains at the right top angle near the punch. The finite elements along the die are found to be distorted especially near the punch, which is evidently caused by friction between the powder and the die wall. This effect is related to the interaction energy of powder particles in the model. The change of the total energy in the powder mixture is shown in figure 3. The results obtained are qualitatively consistent with the works of the authors [2, 9].

Knowing the volume plastic strain, the relative powder density can be calculated via $\rho = \rho_0 \cdot \varepsilon_v^{pl}$.

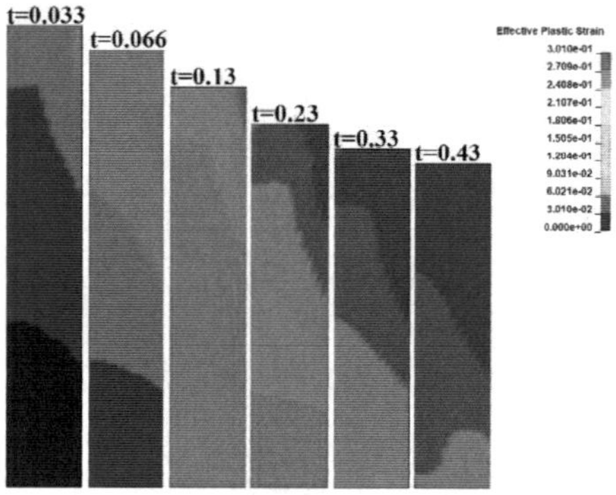

Figure 2. Effective plastic strain in the model during compaction.

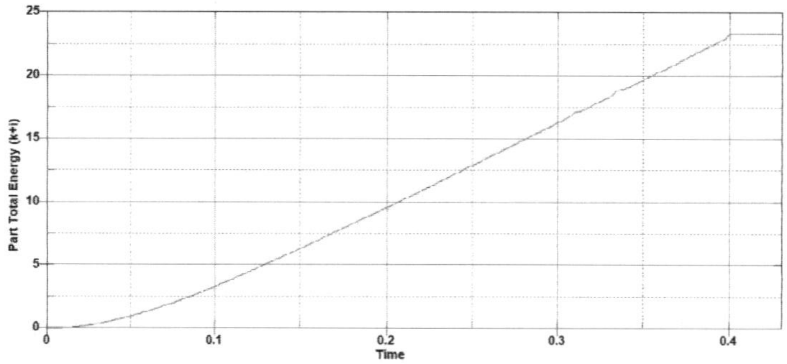

Figure 3. Total energy change during compaction.

Acknowledgments
The results were obtained in the framework of the Integrated Project “Establishment of import-substituting high-tech full-cycle production of complex-shaped indexable carbide cutting inserts for priority industries” (Agreement No. 075-11-2019-036 dated November 27, 2019) implemented by the ISPMS SB RAS and/or TSU at the financial support of the Ministry of Education and Science of the Russian Federation as part of Decree of the Government of the Russian Federation No. 218 dated April 09, 2010.

References

[1] Majzoobi G H and Jannesari S 2015 *Adv. Powder Technol.* **26** 928–934
[2] Lee S C and Kim K T 2002 *Int. J. Mech. Sci.* **44** 1295–1308
[3] Selig S G and Doman D A 2015 *J. Appl. Math.* **2015** 283843
[4] Krok A, Peciar M, Fekete R 2014 *Particuology* **16** 116–131
[5] Andersson D C, Lindskog P, Staf H and Larsson P-L 2014 *J. Mater. Eng. Perform.* **23** 2199–2208
[6] Han L H, Elliott J A, Bentham A C, Mills A, Amidon G E and Hancock B C 2008 *Int. J. Solids Struct.* **45** 3088–3106
[7] Hayashi Y, Otoguro S, Miura T, Onuki Y, Obata Y and Takayama K 2014 *Chem. Pharm. Bull.* **62**(11) 1062–1072
[8] LS-DYNA 2017 *Keyword user's manual* vol II *Material Models* (Livermore: Livermore Software Technology Corporation)
[9] Häggblad H-Å 1985 *Constitutive modelling of hard metal powder* Lic. Thesis 003L (Luleå: Luleå Univ. of Technology) 111

Influence of the angle interaction of the projectile with the wire mesh

L R Akhmetshin[1,2], A G Kushnarev[2] and S V Pashkov[2]

[1]Institute of Strength Physics and Materials Science of Siberian Branch of Russian Academy of Sciences, 2/4 Akademicheskii pr., Tomsk, 634055, Russia
[2]Tomsk State University, 36 Lenin ave., Tomsk, 634050, Russia

E-mail: akhmetshin.lr@gmail.com

Abstract. The problem of interaction of the high-velocity solid spherical shock element simulating a micrometeoroid and the combined protective screen including the mesh barrier is considered. The mesh layer is made of steel wire by weaving method. As a result of numerical modeling by SPH method the results of protective wire mesh and projectile destruction behavior are obtained. The effectiveness of several variants of multilayer screen of equal specific mass with different angles of mesh layer orientation is compared.

1. Introduction

The interaction of high-velocity a micrometeoroid mimic projectiles with protective screens has been a subject of research for quite a long time. Increase of efficiency of antimeteoroid protection of spacecraft is connected with optimal correlation of mass and reliability, as weighting of spacecraft structure, as a rule, is inadmissible because of a number of restrictions, which are laid down during its design.

The use of metal wire mesh as the elements of combined protective screen allowing to reduce the total weight of spacecraft protective shield while maintaining its efficiency looks promising [1]. Special effectiveness is demonstrated by protective screens in the form of corrugated metal mesh, in which the degree of crushing of a projectile can increase more than twice [2], due to the effect "grater" on the inclined wire mesh and increase of interaction time.

The main purpose of this work was to estimate the influence of corrugation mesh angle on the efficiency of conditional corrugated mesh screen [2] at interaction of micrometeoroid with combined protective screen.

2. Mathematical statement

2.1. Mathematical model

Numerical modeling in three-dimensional geometry was performed on the basis of the complete system of equations of mechanics of deformable solids by the method of smoothed-particle hydrodynamics (SPH) using the license package LS-DYNA. The absence of a finite element mesh allows the SPH method to naturally simulate the fragmentation of interacting elements and the motion of the fragments group at a high-velocity impact, and a good correlation with natural tests is noted [3].

In most high-speed experiments, a body made of Al-alloy is used as of a particle acting as a projectile, because it is close in size and density to chondrites, the most common in Earth orbit micrometeoroids.

Therefore, the materials used in this work are aluminum alloy – projectile, one layers of bumper, steel – corrugated metal mesh. To describe the mechanical behavior of the projectile and the wire mesh barrier at high-speed interaction, the Mie–Gruneisen's equation of state as a function of particle velocity $\upsilon_s(\upsilon_p)$ defines the pressure for compressed materials, and the Johnson–Cook model is taken as the defining relations describing the behavior of an elastoplastic body. The similar model used in this work is found in the works of other authors [4-7]. Verification of the model is confirmed in [8, 9].

The physical-mechanical parameters and strength parameters of the Johnson–Cook model for aluminum alloy and steel used in the calculations are given in table 1; here ρ_0 is the initial density of the material, G is the shear modulus.

Table 1. The mechanical constants of the material.

	ρ_0 (kg/m^3)	G (GPa)	C	S_1	γ_0	a
Projectile, Bumper	2770	27.6	5328	1.338	2	0
Grid	7750	81.8	4569	1.49	2.17	0
	A (Pa)	**B** (Pa)	**C**	**n**	**m**	**T**$_m$(K)
Projectile, Bumper	$3.37 \cdot 10^8$	$3.43 \cdot 10^8$	0.01	0.41	1	603.85°
Grid	$1.54 \cdot 10^9$	$4.77 \cdot 10^8$	0.012	0.18	1	1489.85°

2.2. Geometry of the model

The high-velocity interaction of the projectile with the multilayer spaced barrier (layer of corrugated steel wire mesh and Al bumper) was carried out normally, at different angles of orientation of the wire mesh layer (figure 1).

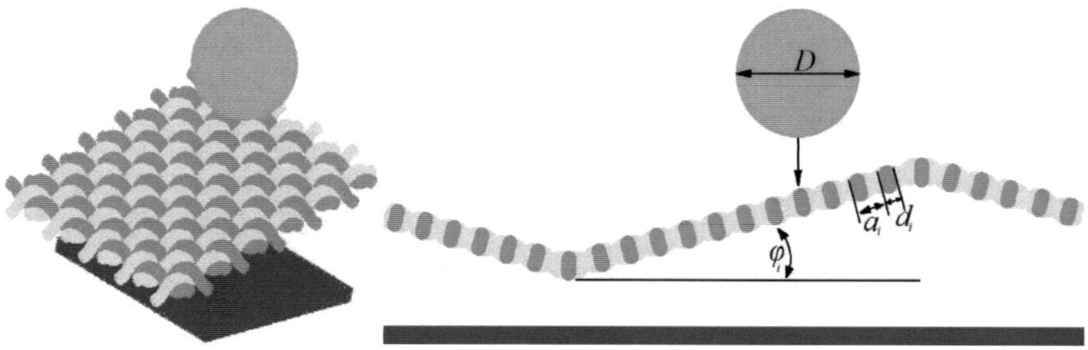

Figure 1. The geometry of the grid and the projectile.

The corrugated mesh layer is modeled by the mesh section at a corresponding slope. At zero corrugation angle φ (angle of mesh layer deviation from the plane of combined protective screen), the corrugated mesh layer is located in the plane of shield. Calculations were made for φ = 15°, 30°, 45°, 60°, 75° (table 2).

Table 2. Initial parameters of the model.

D (mm)	a (mm)	d_{15} (mm)	d_{30} (mm)	d_{45} (mm)	d_{60} (mm)	d_{75} (mm)
2.3	0.556	0.244	0.231	0.209	0.175	0.126

Comparison of efficiency of different protective screens makes sense only under condition of their mass coincidence. Therefore, when the corrugation angle (i.e. the angle of mesh inclination) is changed, the diameter of the wires is recalculated according to the equation in order to preserve the specific density (in the plane of the screen, perpendicular to the direction of impact):

$$d = d_0 \cdot \sqrt{\cos(\varphi)}$$

where φ is the angle between the wire mesh and the horizontal axis, d_0 is the wire diameter at $\varphi = 0^o$.

2.3. Boundary condition

To simulate the high-velocity interaction of the projectile with the mesh barrier, the boundary conditions were applied as follows, on the lower (rigid fixation) and upper (movement in the Y axis and free movement in the X and Z axes) boundaries:

$$U_x^{P_{KLMN}} = U_y^{P_{KLMN}} = U_z^{P_{KLMN}} = 0(m)$$

$$V^{proj} = 5000\left(\frac{m}{s}\right)$$

where $U_{x,y,z}^{P_{KLMN}}$ the rigid closing along the perimeter of the mesh barrier and bumper, V^{proj} the projectile velocity is normal to the bumper, imitating the shell of the spacecraft.

3. Results and discussion

Calculations show (figure 2) that with a small angle of corrugation, the destruction of the projectile starts from its front part immediately after contact and is characterized by the formation of jets (high-velocity fragmentation flow streams) ejected in the direction of impact. The jets formed after the impact are not ejected normally, but disperse at some angle, which in turn will increase the area of impact. This is the main feature of fragmentation when interacting with the wire grid barrier. The number of jets correlates with the number of cells hitting the projectile. The energy of the moving particles is reduced by active mass scattering of the fragments.

Figure 2. Destruction of projectile before interaction with bumper at $\varphi = 15^o$.

As the angle φ increases (figure 3), the jets become less pronounced, the effect of "grater" of the projectile on the wire mesh increases, while the speed of the outgoing jets continues to be predominantly directed along the normal to the mesh bumper. Thus, the most part of the fragment cloud deviates from the impact direction and changes the barrier impulse. This leads to a decrease in the penetration ability of the projectile, in particular, for this task decreases the size of the hole punched in the bumper. Additionally, an increase in the mass of fragments sliding along the mesh layer can be noted, up to the rebound of the unbroken part at large corrugation angles.

Theoretical estimation shows that high-velocity interaction of an aluminum projectile with the corrugated metal mesh as well as with layers of bumpers can lead to sufficient heating to the melting point. Obviously, in experiments, heating should lead to a sharp drop in the resistance of the projectile material.

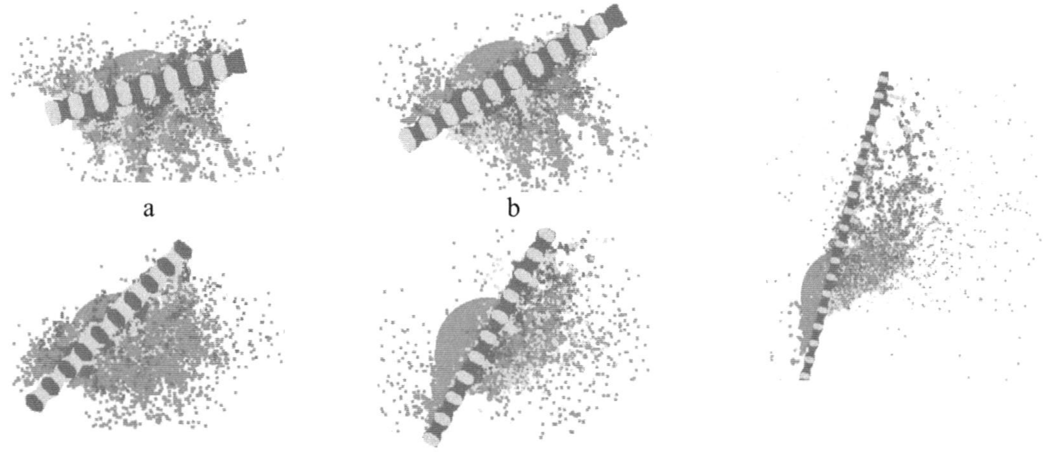

a b c d e

Figure 3. High-speed interaction of the projectile with the wire mesh, a) 15°, b) 30°, c) 55°, d) 60°, e) 75°. Side view.

4. Conclusion

In the work problem of interaction of high-velocity micrometeoroid with the combined protective screen including the layer from the corrugated wire mesh was numerically solved. The mechanism of interaction and destruction of projectile–mesh system at different corrugation angles was investigated. Variability of specific weight of protective screen for different corrugation angles was provided by change of wire diameter of the mesh.

At small angles of deviation from normal interaction, the wire mesh barrier is destroyed as a thin-walled sheet of the continuum material. With an increase in the angle of inclination of the mesh layer, the effect of grater of the projectile against the mesh increases, with the fragmentation cloud shifting away from the impact axis. The share of the projectile that ricochet along the wire mesh also increases, and for smaller projectile it is possible to rebound the undestroyed part of the projectile.

Acknowledgments
This study is funded by the Russian Scientific Foundation, project no. 16-19-10264.

References.

[1] Myagkov N N and Shumikhin T A 2019 *AIMS Materials Science* **6**(5) 685-696
[2] Dobritsa D B, Pashkov S V and Khristenko Yu F 2020 *Cosmic Research* **58**(2), 105-110
[3] Becker M, Seidl M, Mehl M, Souli M and Legendre J-F 2019 *12th European LS-DYNA Conference*, Koblenz, Germany
[4] Myagkov N N, Shumikhin T A and Ponomarev 2017 *Mechanics of composite materials and structures* **22**(3) 452-463
[5] Myagkov N N, Shumikhin T A and Bezrukov L N 2010 *Int J Impact Eng* **37** 980-994
[6] Myagkov N N, Silant'ev M V 2013 *Mechanics of composite materials and structures* **19**(4) 475-491
[7] Myagkov N N 2017 *J Exp Theor Phys* **124** 57-69
[8] Akhmetshin L R, Moskvichev E N and Skripnyak V V 2019 *IOP Conf Ser: Mater Sci Eng* **511** 012013
[9] Nikhamkin M Sh, Voronov LV, Lyubchik O L and Gladkiy I L 2011 Izvestiya Samarskogo nauchnogo centra RAN **13** (4) 991-997

Antibacterial PLGA and PCL membranes, modified by magnetron sputtering method of copper target

A D Badaraev, D V Sidelev and S I Tverdokhlebov

Tomsk Polytechnic University, 30 Lenin Avenue, 634050, Tomsk, Russia

E-mail: tverd@tpu.ru

Abstract. The paper discusses barrier membranes which were formed from poly(lactide-co-glycolide) and polycaprolactone by electrospinning method. Then, plasma modification of membranes was carried out by magnetron sputtering of copper target in argon atmosphere. The morphology, elemental composition, mechanical and antibacterial properties of the obtained samples were investigated. Modification with copper in the magnetron discharge plasma allows maintaining the original characteristics of pristine polymeric samples and imparting antibacterial properties to them.

1. Introduction

Periodontitis is a common inflammatory disease of the periodontal tissue, which leads to jaw bone resorption and teeth loss. Barrier membranes are often used to treat such disease, which main function is preventing the penetration of fast-growing soft tissue into osseous defects in the alveolar bone and stimulation of proliferation the bone tissue in them [1]. For periodontitis treatment with barrier membranes, the guided tissue/bone regeneration (GTR/GBR) procedure is required. Nowadays, GTR/GBR membranes consist of natural and synthetic materials based on biocompatible and biodegradable polymers. Natural polymers have a high biocompatibility, but has excessively high degradation rate and low strength characteristics [2]. This can lead to membrane failure, which require additional surgical procedures. Compared to natural polymers, synthetic membranes have higher strength characteristics and longer degradation times [2]. Most commonly used biodegradable synthetic polymer for barrier membrane manufacturing is poly(lactide-co-glycolide) (PLGA) [3]. That polymer consists of lactic and glycolic acids, which allows PLGA to have a controllable degradation time [4]. Polycaprolactone is biodegradable aliphatic polyester with extremely high elongation at break (PCL) [5]. Therefore, PCL is used like a material in different types of medical devices including barrier membranes [6].

One of the most common methods for the manufacture of polymer membranes is electrospinning, since it allows the formation of highly porous scaffolds with a developed structure, which have the ability to mimic the extracellular matrix [7].

Nevertheless, a synthetic scaffold obtained by electrospinning method doesn't have antibacterial properties, which can increase chances of inflammatory diseases in the case of their introduction into living organism. Previously we have shown, that polymer membranes modified by magnetron sputtering of copper target suppress the growth of bacteria St. Aureus [8]. Relatively low temperature of magnetron sputtering process allows modify the surface of non-woven polymers saving the original morphology and structure without burn-through. Comparing with atmospheric plasma jet and chemical vapor deposition, with magnetron sputtering aligned high-purity thin films can be created. Magnetron plasma modification of PLGA and PCL membranes by copper could provide the antibacterial properties to them, which could decrease the probability of surgical complications caused by the infection penetration into oral cavity.

The aim of this work is development of PLGA and PCL barrier membranes modified by copper to possess them antibacterial properties for further use in periodontitis treatment.

2. Materials and methods

2.1. Barrier membranes formation

Barrier membranes were made from 4 % PLGA (85/15, Corbion Purac, Netherlands) solution in hexafluoroisopropanol ($(CF_3)_2CHOH$, P&M Invest, Russia) and 8 % PCL (Sigma-Aldrich, UK) solution in chloroform ($CHCl_3$, Fisher Scientific, UK). Electrospinning of polymeric solutions was provided on installation NANON-01A (MECC Co, Japan) with regimes for PLGA membranes manufacturing: cylindrical collector 200 mm long and 100 mm diameter; solution flow rate – 4 ml/h; voltage between needle and collector – 22 kV; distance between needle and collector – 150 mm; collector rotational speed – 200 rpm. For PCL membranes, regimes were the same except: distance between needle and collector – 140 mm; collector rotational speed – 50 rpm.

2.2. Barrier membranes modification

Membranes plasma modification by magnetron sputtering of copper target (Cu – 99.95 %) was carried out using an ion-plasma installation [9] with direct current power supply (APEL-M-5PDC, JSC Apelvac, Russia) under the following technological conditions: discharge power – 600 W; current – 1.1 A; deposition time – 15.7 min; operation pressure in the chamber – 0.3 Pa; gas – argon (99.99 %).

2.3. Investigation of barrier membranes

SEM-images with 1000× magnification were obtained by scanning electron microscope (SEM) JCM-6000Plus (Jeol, Japan). Elemental composition of samples was provided by method of energy dispersive X-ray spectroscopy (EDAX). The fiber diameter and porosity measurements of PLGA and PCL membranes were carried out in the ImageJ 1.48 software (National Institutes of Health, USA), using the plug-in Diameter J v1.018 (National Institute of Standards and Technology, USA).

The strength characteristics of the membranes were evaluated using an Instron 3343 instrument (Illinois Tool Works, USA) equipped with an Instron 2519-102 sensor (Illinois Tool Works, USA).

The antibacterial activity was assessed in accordance with ISO 20732: 2013. Suspension with St. Aureus (ATCC 25923) was applied to the surface of 10 x 10 mm specimens.

The statistical analysis of the obtained data was performed using Mann-Whitney and One-way ANOVA tests in OriginPro v.2019b (OriginLab, USA) software. Differences were considered statistically significant at $p < 0.05$.

3. Results

3.1. Scanning electron microscopy and EDAX

SEM images, fiber diameter distribution histograms and EDAX spectra of PLGA and PCL membranes are shown in figure 1.

On SEM images are shown, that barrier membranes consist of polymeric fibres, which forms chaotically interlaced nonwoven structure. The EDAX spectra of control samples shows three reflections corresponding to carbon (C), oxygen (O), gold (Au), last of which was deposited to increase electrical conductivity. Plasma modified membranes have a high intensity copper (Cu) peaks on EDAX spectra.

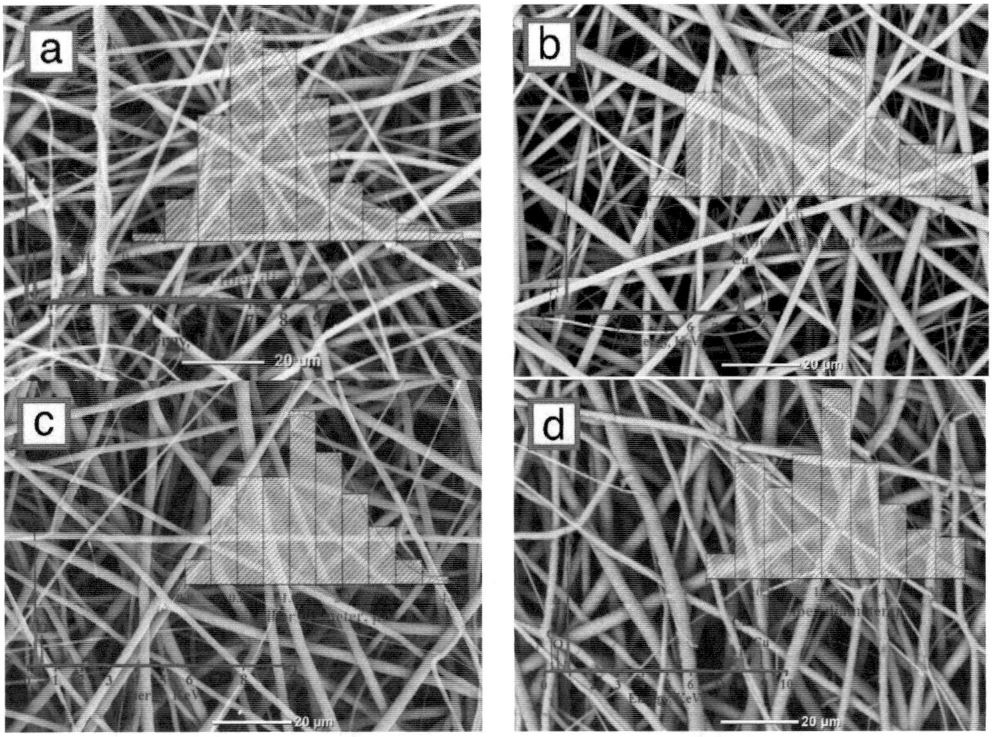

Figure 1. SEM images, fiber diameter histogram and EDAX spectra of:
a) PLGA (pristine sample); b) Cu-PLGA (modified sample);
c) PCL (pristine sample); d) Cu-PCL (modified sample).

Mean fiber diameters and porosity of PLGA and PCL membranes are shown in table 1.

Table 1. Values PLGA and PCL membranes fiber diameters and porosity.

Sample names	Fiber diameter (μm)	Porosity (%)
PLGA	1.82±0.63	61±5
Cu-PLGA	1.72±0.74	57±5
PCL	1.75±0.77	56±3
Cu-PCL	1.81±0.73	54±4

Modification of PLGA and PCL membranes with copper does not significantly affect on the fiber diameters and porosity. The retention of the average fiber diameter and porosity after plasma modification indicates that the modification under the selected modes does not affect the surface morphology of PLGA and PCL membranes.

3.2. Mechanical properties

Values of the membranes mechanical characteristics are shown in table 2.

Table 2. Mechanical properties of PLGA and PCL membranes.

Sample names	Tensile strength, MPa	Relative elongation, %	Young's modulus, MPa
PLGA	3.3±0.3	260±25	80±10
Cu-PLGA	3.5±0.1	230±10	95±1
PCL	3.2±0.3	850±50	6.4±0.3
Cu-PCL	3.3±0.2	820±20	7.5±0.2

Pristine PLGA and PCL membranes have a comparable values of tensile strength. PCL membrane has a ~ 3 times higher relative elongation than PLGA.

After plasma modification of PLGA and PCL membranes, values of tensile strength, elongation and Young's modulus do not change reliably. This is primarily due to the retention of the morphology and surface structure of membranes after their modification.

3.3. Antibacterial activity

Indicators of the antibacterial activity of membranes are shown in table 3.

Table 3. Antibacterial properties of PLGA and PCL membranes.

Sample names	**Bacteria's number (0 h), CFU×10^4/ml**	**Bacteria's number (24 h), CFU×10^7/ml**	**Antibacterial activity (A)**
PLGA	1.9±0.4	6.0±1.0	-
Cu-PLGA	1.8±0.2	0.2±0.1	1.51
PCL	1.3±0.2	10.6±1.3	-
Cu-PCL	1.6±0.4	0.3±0.1	1.44

PLGA and PCL membranes, modified with copper have a significant antibacterial activity (A> 1.0). In comparison with the number of bacteria on pristine samples after 24 hours, the colony forming units (CFU) on modified samples have decreased by ~ 97%.

4. Summary

Modification by the method of magnetron sputtering of a copper target makes it possible to impart significant antibacterial properties to barrier membranes while maintaining the surface morphology and their mechanical properties. In this way magnetron modification of PLGA and PCL materials by copper can be recommended for creating barrier membranes with ability of effective prevention of infectious complications in the oral cavity.

Acknowledgments

This research was financially supported by Tomsk Polytechnic University Competitiveness Enhancement Program project VIU-SEC B.P. Veinberg-196/2020.

We express our gratitude to A L Nemoykina, employee of laboratory of biopolymers and biotechnology in National Research Tomsk State university, for conducting research of antibacterial properties

References

[1] Herrera D et al 2008 Subgingival microbial profiles in chronic periodontitis patients from Chile *Journal of clinical periodontology* **35** 106-113

[2] Gentile P et al 2011 Polymeric membranes for guided bone regeneration *Biotechnology journal* **6** 1187-1197

[3] Houchin M, Topp E 2008 Chemical degradation of peptides and proteins in PLGA: a review of reactions and mechanisms *Journal of pharmaceutical sciences* **97** 2395-2404

[4] Gentile P et al 2014 An overview of poly (lactic-co-glycolic) acid (PLGA)-based biomaterials for bone tissue engineering *International journal of molecular sciences* **15** 3640-3659

[5] Nottelet B et al 2009 Factorial design optimization and in vivo feasibility of poly (ε-caprolactone)-micro-and nanofiber-based small diameter vascular grafts *Journal of Biomedical Materials Research Part A* **89** 865-875

[6] Yang F et al 2009 Development of an electrospun nano-apatite/PCL composite membrane for GTR/GBR application *Acta biomaterialia* **9** 3295-3304

[7] Han D, Gouma P 2006 Electrospun bioscaffolds that mimic the topology of extracellular matrix *Nanomedicine : Nanotechnology, Biology, and Medicine* **2** 37-41

[8] Badaraev A et al 2020 Piezoelectric polymer membranes with thin antibacterial coating for the regeneration of oral mucosa *Applied Surface Science* **504** 144068
[9] Sidelev D et al 2017 A comparative study on the properties of chromium coatings deposited by magnetron sputtering with hot and cooled target *Vacuum* **143** 479-485

Formation of SERS-substrate based on Ag-Cu bimetallic nanoparticles for detection of ultra-small amounts of substance

S Dubkov[1], A Savitskiy[1,2], A Tarasov[1], D Novikov[1], N Mineeva[1], M Bazarova[2], A Overchenko[1] and D Gromov[1]

[1]Institute of Advanced Materials and Technologies, National Research University of Electronic Technology (MIET), bld. 1, Shokin Square, Zelenograd, Moscow, Russia, 124498
[2]Scientific-Manufacturing Complex «Technological Centre», House 1, Bld. 7, Shokin Square, Zelenograd, Moscow, Russia, 124498

E-mail: sv.dubkov@gmail.com

Abstract. The aim of this work was to determine the sensitivity threshold and enhancement factor of a planar SERS substrate based on an array of composite Ag - Cu nanoparticles. The nanoparticles were deposited using the method of vacuum-thermal evaporation followed by annealing at 300 ° C. Methylene blue was chosen as the analyte. The possibility of detecting the micro- and nanomolar concentration of methylene blue using a SERS-substrate with an active Ag-Cu layer at a laser wavelength of 632.8 nm is shown. The use of such arrays of nanoparticles as an active layer makes it possible to achieve an analytical enhancement factor of the SERS substrate of the order of 6×10^5.

1. Introduction
Raman spectroscopy is a common chemical analysis technique that provides information on the composition and structure of a test substance. The presence of plasmonic nanostructures near the analyte surface can lead to significant enhancements (up to $\sim 10^{12}$) of the Raman signal of the substances under study. The effect of surface-enhanced Raman scattering (SERS) of light arises due to the excitation of plasmon resonance in metallic nanostructures. SERS can be used to detect substances of low concentration, which makes it a promising method for such areas as materials science (determination of the composition and structure of nanoscale objects), pharmacology (control of the composition of dosage forms), forensics (determination of explosives or drugs), early diagnosis of many diseases, including cardiovascular and viral, as well as malignant tumors [1-8]. Due to their unique optical properties, Ag, Au, and Cu [9] and their alloys [10] are traditional materials used in the formation of SERS substrates.

In our previous work [11], using thin layers of amorphous carbon as an analyte as an example, we showed that the use of composite Ag-Cu arrays for SERS allows a high amplification of the Raman signal in the red spectral region, in which copper arrays are usually effective. However, the enhancement for Ag-Cu turns out to be higher than that for arrays of pure copper particles: it turns out to be comparable with the enhancement of an array of pure Ag nanoparticles in the blue spectral region. In this work, using planar SERS substrates based on arrays of composite Ag-Cu nanoparticles, a solution of methylene blue (MB) dye at various concentrations (normal, mil-, micro-, nanomolar), which is a standard marker for assessing the gain, was studied.

2. Experimental

SERS substrates were planar structures on silicon containing a reflective metal layer (Al), a thin dielectric layer (SiO_2), and an SERS-active layer (Ag-Cu NPs). The manufacturing process of SERS substrates and their cross section are schematically shown in Figure 1.

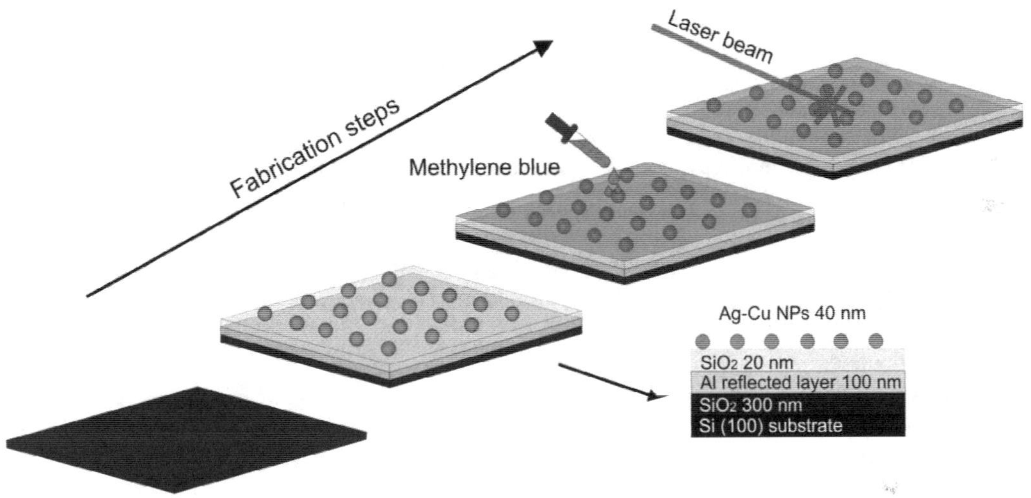

Figure 1. Fabrication steps of SERS-substrate based on Ag-Cu alloy nanoparticles.

Arrays of Ag-Cu nanoparticles were formed by vacuum-thermal evaporation and condensation of a substance on substrates, followed by thermal annealing in vacuum at 300 ° C. In this case, the ratio of the weight portions of the evaporated metals was chosen so that the formed condensate corresponded to the eutectic composition (Ag - 5.0 mg; Cu - 2.0 mg). Studies in [11] show that the particles obtained in this way are precisely composite and consist of parts of Cu and Ag.

In this series of experiments, we used a LabRAM HR Evolution Raman spectrometer (Horiba). The measurements were carried out at room temperature at a wavelength of 632.8 nm. The nominal laser power was 10 mW. Methylene blue (Sigma-Aldrich, CAS Number 122965-43-9) converted into solutions of nano-, micro-, milli- and normal concentrations was used as an analyte. 10 μL of the analyte was applied to a substrate by pipette, followed by drying the samples in an oven at room temperature.

3. Results and discussion

At the first stage of the study of methylene blue using a SERS-substrate with Ag-Cu nanoparticles, an experiment was carried out to identify the optimal laser power for obtaining the Raman spectra of the analyte. Figure 2 shows the spectra of MB obtained at a laser power of 0.1 to 10% of its nominal power. At 10% power, the average intensity is about 11300 arbitrary units. For a power of 5%, the average intensity is 8100 a.u. For a power of 2.5%, the average intensity is approximately 4650 a.u. For a power of 1%, the average spectrum intensity is 2200 a.u. For a power of 0.1%, the average spectrum intensity is about 160 a.u.

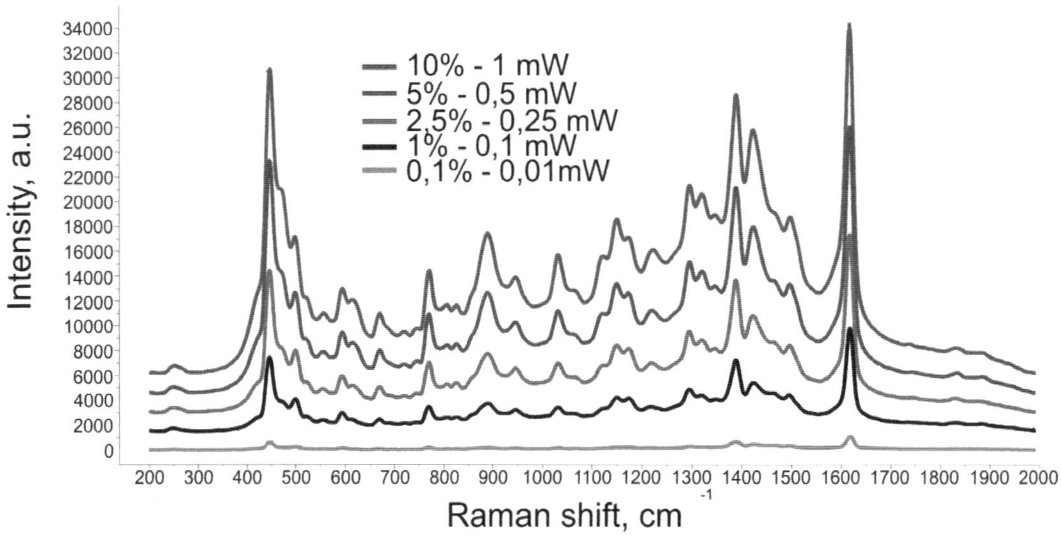

Figure 2. Raman spectra for SERS substrate based on Ag-Cu MB millimolar concentration at laser power 0.1%, 1%, 2.5%, 5%, 10% of its nominal power.

As can be seen from Figure 2, with decreasing laser power, the intensity of the spectra decreases. It is important to note that when the laser power is increased, the formation of new peaks does not occur, which may appear due to the thermal effect of laser radiation on the analyte. The revealed spectral modes of MB millimolar concentration (Figure 2) using the developed SERS substrate are in full agreement with the measurement data of other authors [12, 13].

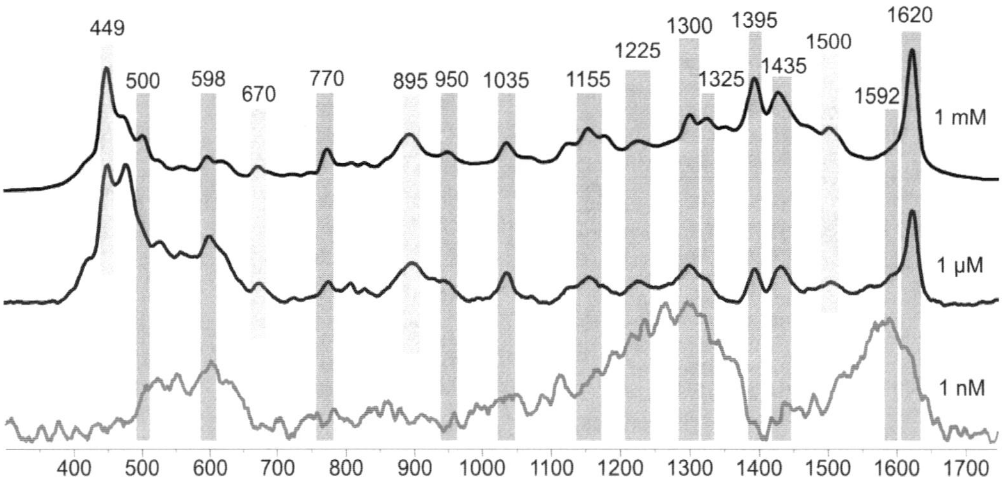

Figure 3. Raman spectra for a SERS substrate based on Ag-Cu methylene of blue mil-, micro- and nanomolar concentrations at a laser wavelength of 632.8 nm.

Figure 3 shows the spectra of MB solutions of 1 milli-, micro-, and nanomolar concentrations obtained at a laser wavelength of 632.8 nm and 1 mW power. It can be noted that lowering the concentration to micromolar does not affect the clarity of the Raman spectra peaks. At the same time, for the nanomolar concentration of MB, the peaks of most of the spectral lines are not clearly detected, while some of them are shifted. However, it should be noted that a part of the spectrum for nanomolar concentration is similar to the spectra of micro- and millimolar concentrations. However, the clear peaks for nanomolar concentration are reduced to 10, while for millimolar and micromolar concentrations are around 17. Figure 4 shows the Raman spectra of MB with micromolar concentration on a glass substrate and on a SERS substrate with an active layer of composite Ag - Cu nanoparticles. It should be noted that in the case of a glass substrate, none of the characteristic MB spectral modes is observed, even for the strong vibrations of MB attributed to the skeletal deformation of C-N-C (at 448 and 500 cm^{-1}) and the ring stretching of C-C (at 1623 cm^{-1}).

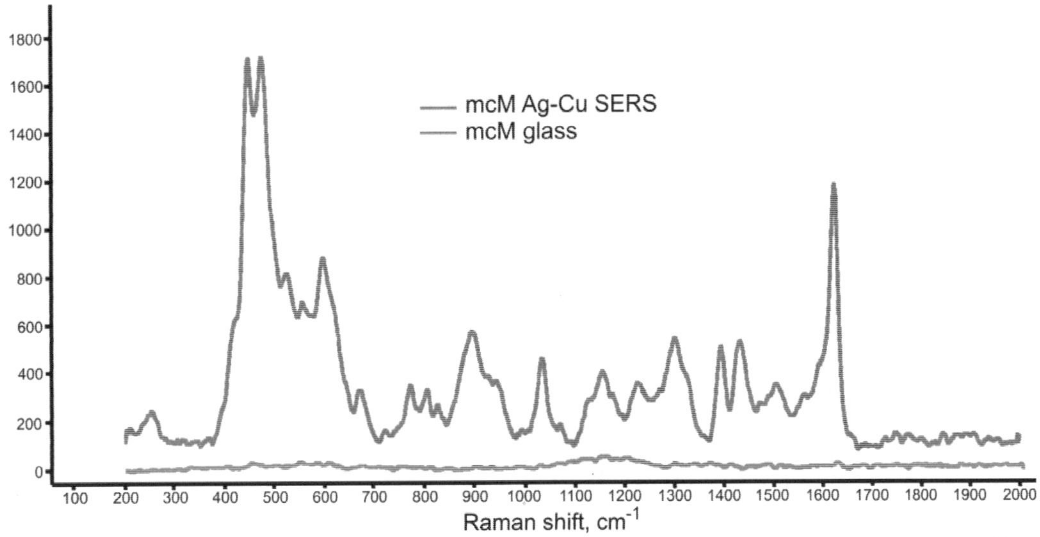

Figure 4. Raman spectra of the micromolar MB concentration on a glass substrate and on a SERS substrate with an active layer of composite Ag-Cu nanoparticles.

The analytical enhancement factors (AEF) calculated for spectral modes at 1623 and 440 cm^{-1} for planar SERS substrates with an active layer based on an array of Ag-Cu nanoparticles for a laser with a wavelength of 632.8 nm was ~ 5.78×10^5 (1623 cm^{-1}) and 5.71×10^5 (440 cm^{-1}). The calculation of the AEF was carried out according to the well-known method [14]. The obtained AEF values turned out to be quite close to the enhancements reported in [12 14], in which the authors studied MB using Au nanoparticles (EF 2.6×10^4) [12] and NPs Ag of various shapes (AEF 3.7×10^5) [14]. The relative standard deviation (RSD) of the Raman intensity at 1623 cm – 1 was about 23%, which is significantly higher than our previous estimates when detecting a thin a – C film (~ 11%) [11]. This difference is probably related to the nonuniform distribution of the surface concentration obtained upon drying of a droplet of micromolar MB concentration, and not to the bulk of Ag - Cu nanoparticles itself. Previous measurements on a-C film are more accurate in terms of evaluating the uniformity of Raman signal amplification by the SERS-active layer, since the analyte distribution over the surface was more uniform.

The enhancement for MB obtained in this work is not one of the best, for example, there are reports on the detection of femtomolar concentrations of MB using structures of complex shape with a developed surface based on Ag – Au nanoparticles [15]. Nevertheless, our result indicates that the use of the Ag - Cu composite in SERS detection is promising. Undoubtedly, the use of nanoobjects of this material of a more complex shape can increase the sensitivity of SERS substrates and limiting quantitative determination. This should become one of the tasks of further research.

4. Conclusions

The work shows the possibility of detecting the micro- and nanomolar concentration of MB using a SERS-substrate based on composite Ag-Cu nanoparticles at a laser wavelength of 633 nm. The use of such arrays of nanoparticles as an active layer makes it possible to achieve an analytical enhancement factor of the SERS substrate of the order of 5.7×10^5. It has been demonstrated that when using different laser powers, no new peaks appear due to the interaction of laser radiation with MB.

Acknowledgements
This work was financially supported by the grant of President of the Russian Federation (project MK-2222.2019.8)

References

[1] Subramaniam S and Fletcher C 2018 Trimethylamine N-oxide: breathe new life: Trimethylamine N-oxide (TMAO): Breathe new life *British Journal of Pharmacology* **175** 1344–53
[2] Kukushkin V I, Ivanov N M, Novoseltseva A A, Gambaryan A S, Yaminsky I V, Kopylov A M and Zavyalova E G 2019 Highly sensitive detection of influenza virus with SERS aptasensor ed K Kourentzi *PLOS ONE* **14** e0216247
[3] Tan Y, Yan B, Xue L, Li Y, Luo X and Ji P 2017 Surface-enhanced Raman spectroscopy of blood serum based on gold nanoparticles for the diagnosis of the oral squamous cell carcinoma *Lipids in Health and Disease* **16**
[4] Gromov D G, Borgardt N I, Volkov R L, Galperin V A, Grishina Y S and Dubkov S V 2013 Specific features of the structure and properties of carbon nanocolumns formed by low-temperature chemical vapor deposition *Semiconductors* **47** 1703–6
[5] Bandarenka H, Artsemyeva K, Redko S, Panarin A, Terekhov S and Bondarenko V 2013 Effect of swirl-like resistivity striations in n $^+$ -type Sb doped Si wafers on the properties of Ag/porous silicon SERS substrates *physica status solidi (c)* **10** 624–7
[6] Bandarenka H, Redko S, Smirnov A, Panarin A, Terekhov S, Nenzi P, Balucani M and Bondarenko V 2012 Nanostructures formed by displacement of porous silicon with copper: from nanoparticles to porous membranes *Nanoscale Research Letters* **7**
[7] Arzumanyan G, Doroshkevich N, Mamatkulov K, Shashkov S, Girel K, Bandarenka H and Borisenko V 2017 Phospholipid detection by surface-enhanced Raman scattering using silvered porous silicon substrates: Phospholipid detection by surface-enhanced Raman scattering *physica status solidi (a)* **214** 1600915
[8] Girel K, Yantcevich E, Arzumanyan G, Doroshkevich N and Bandarenka H 2016 Detection of DNA molecules by SERS spectroscopy with silvered porous silicon as an active substrate: Detection of DNA molecules by SERS spectroscopy *physica status solidi (a)* **213** 2911–5
[9] Lee J, Min K, Kim Y and Yu H K 2019 Surface-Enhanced Raman Spectroscopy (SERS) Study Using Oblique Angle Deposition of Ag Using Different Substrates *Materials* **12** 1581
[10] Fan M, Lai F-J, Chou H-L, Lu W-T, Hwang B-J and Brolo A G 2013 Surface-enhanced Raman scattering (SERS) from Au:Ag bimetallic nanoparticles: the effect of the molecular probe *Chem. Sci.* **4** 509–15
[11] Dubkov S V, Savitskiy A I, Trifonov A Y, Yeritsyan G S, Shaman Y P, Kitsyuk E P, Tarasov A, Shtyka O, Ciesielski R and Gromov D G 2020 SERS in red spectrum region through array of Ag–Cu composite nanoparticles formed by vacuum-thermal evaporation *Optical Materials: X* **7** 100055
[12] Li C, Huang Y, Lai K, Rasco B A and Fan Y 2016 Analysis of trace methylene blue in fish muscles using ultra-sensitive surface-enhanced Raman spectroscopy *Food Control* **65** 99–105
[13] Xiao G-N and Man S-Q 2007 Surface-enhanced Raman scattering of methylene blue adsorbed on cap-shaped silver nanoparticles *Chemical Physics Letters* **447** 305–9
[14] Zannotti M, Rossi A and Giovannetti R 2020 SERS Activity of Silver Nanosphere, Triangular Nanoplates, Hexagonal Nanoplates and Quasi-Spherical Nanoparticles: E ect of Shape and Morphology *Coatings* **10** 228
[15] William R V, Das G M, Dantham V R, Laha R 2019 Enhancement of Single Molecule Raman Scattering using Sprouted Potato Shaped Bimetallic Nanoparticles *Scientific Reports* **9** 10771

Investigation of the tangential discontinuities formation in the glycerol under shear load

D V Efremov[1], V V Chudinov[1] and S V Uvarov[2]

[1]Perm Federal Research Center UB RAS, 13a Lenina Street, Perm, 614990, Russia
[2]Perm State University, 15 Bukireva Street, Perm, 614990, Russia

E-mail: efremov.d@icmm.ru

Abstract. The work is devoted to the study of the formation of tangential discontinuities of glycerol parameters under shear load. The studies were carried out using a modified coaxial cylinder viscometer. The experiment was inspired by the work of Ya. I. Frenkel, in which it was noted that "... liquids, being condensed media, should exhibit effects that are more characteristic of solids than gases ...", while the thermodynamic and kinetic properties should be determined by structural "variables", traditional for both "liquid" and solid state. These studies will further help in explaining the anomalous momentum transfer regimes caused by the mechanisms of instability that are realized under the conditions of the "decay" of multiple metastable states, usually observed under intense (dynamic and shock-wave) loading in liquids and solids.

1. Introduction

In experimental and theoretical works of the authors carried out at the ICMM UB RAS inspired by the classic works of Ya.I. Frenkel [1], A.D. Sakharov [2], it was shown for the first time that the existence of shear elasticity of a "close order" in liquids presupposes the formation of mesoscopic defects, shears caused by the coordinated movement of groups of molecules, and the implementation of quasi-plastic mechanisms of momentum transfer associated with collective effects in ensembles of microshears. These effects lead to qualitatively new mechanisms of viscosity and dissipation, the values of which reach asymptotically minimum values due to the multiscale "strong" interactions characteristic of nonequilibrium "critical systems". The possibility of implementing these mechanisms in liquids under less extreme conditions, including in turbulence processes, was first substantiated in the works [3-5].

As a rule, a hydrodynamic approximation is used to describe the flow of a liquid, for example, the Navier-Stokes equation, where the main parameter characterizing the flow is viscosity. At the same time, liquids have the properties inherent in solids - density, modulus of elasticity, heat capacity [6]. Frenkel also noticed the important property of liquids to support high-frequency shear waves of a solid.

Frenkel suggested that liquid particles vibrate for some time, similar to solids, and then diffusely move to neighboring quasi-equilibrium positions [1]. These jumps enable liquid flow and endow liquids with viscosity. He introduced τ as the average time between diffusive jumps and predicted that liquids behave like solids and hence support propagating shear modes at time shorter than τ, or frequency above the Frenkel frequency ω_F:

$$\omega > \omega_F = \frac{1}{\tau}$$

In the work [7] devoted to the study of the flow of metal glasses from the Vitreloy-1 alloy, the formation of shear bands in a deeply supercooled liquid was established. As a result of experimental studies, the authors [7] have shown that the Newtonian flow of melts of metal glasses is realized at high temperatures and low shear rates. A non-Newtonian flow is formed at moderate temperatures and high shear rates; under these conditions, the flow becomes extremely sensitive to the shear rate and, as a consequence, is unstable, which leads to the formation of shear bands.

It is generally accepted that the so-called Newtonian fluids do not have shear strength, and their viscosity does not depend on the flow rate. However, studies of the behavior of liquid media subjected to ultra-weak shear loading show that these media have a certain shear strength.

In [8,9], using an optical technique, it was shown that polar liquids (water, acetone, ethyl alcohol, glycerol, etc.) at rest have an ordered structure, which is destroyed under weak shear deformations. It was shown in [10,11] that glycerin exhibits non-Newtonian properties in the temperature range 223-313 K: "At 313 K, glycerol is a moderately viscous liquid, explaining the moderate damping in the propagation direction. At 273 K, due to coupling with structural relaxation, the glycerol acoustic modes are very strongly damped. The asymmetric character of the wave profile with respect to the interface is linked to the finite both real and imaginary parts of the shear stiffness of the liquid. At 223 K , glycerol has very solidlike properties and low acoustic damping, explaining the lower damping". Taking this into account, it seems important to investigate the possibility of the formation of tangential discontinuities in the parameters in liquid media, which can serve as the beginning of the process of dividing the loaded sample into fragments.

Tangential discontinuities of condensed matter parameters are divided into two types. Tangential rupture of the first degree is the formation of a zone elongated in the direction of shear deformation, where the condition $\mu_z << \mu$ is satisfied, where μ, μz are the values of the structural viscosity of the medium (its definition is given in [12]) outside the zone and in the zone, respectively; the density of the medium in the zone of tangential discontinuity of the first degree and outside it is practically the same. Tangential discontinuity of the second degree is the formation of a zone elongated in the direction of shear deformation in which the conditions $\rho_z / \rho_0 \to 0$, $\mu_z / \mu \to 0$ are satisfied, where ρ_z, ρ_0 are the densities in the zone of tangential discontinuity of the second degree and an undisturbed medium, respectively [13].

2. Experimental part

The formation of tangential discontinuities of the first degree was investigated on an original viscometer with a system of coaxial cylinders (rotary rheometer) [14], which allows not only measuring the dynamic viscosity of the liquid under study but also conducting visual observations and examining liquids with solid particles immersed in them. Rotational viscometers make it possible to measure the true or absolute viscosity of both Newtonian and non-Newtonian (structured or rheological) media, the schematic diagram of the viscometer is shown in Figure 1. The space between two coaxial cylinders 1 and 2 is filled with a liquid medium 3. The stepper motor 4 drives the external cylinder with adjustable linear speed from 0.005 to 6 cm / s, the inner cylinder remains stationary. As a result of the rotation of the outer cylinder, a transition layer is formed in the medium under study, in which, at certain rates, shear deformation modes can exist, which ensure the destruction of the ordering of the structure of the medium. To register the process, the system of coaxial cylinders was placed in the light flux formed by the source 5, between two polarizers 6, after the polarizer, the light flux fell on a photo or video camera 7 recording the process under study. In addition, the inner cylinder 2 is connected to a special elastic element, the deformation of which under the action of viscous stresses in the liquid under study is measured using a high-precision laser range finder. The external view of the experimental setup is shown in Figure 2.

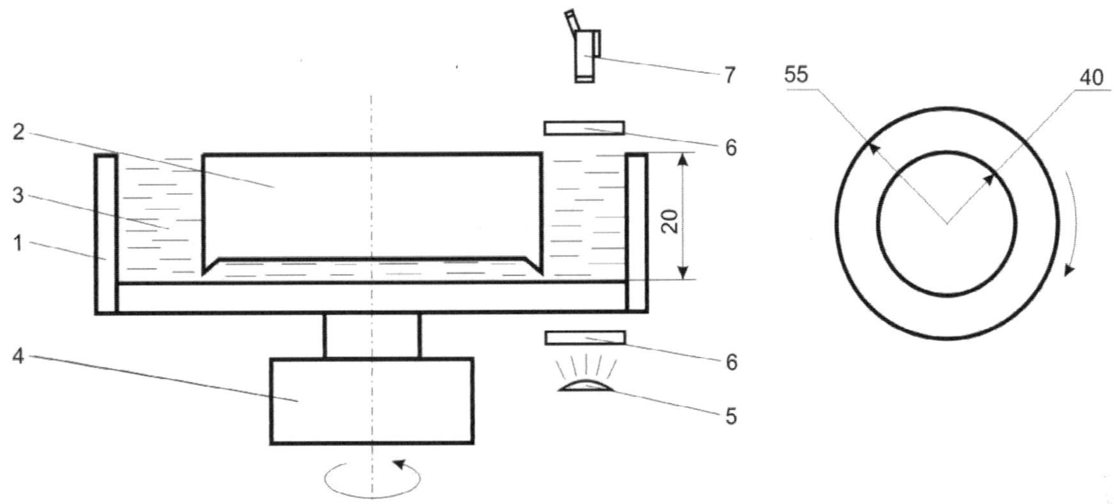

Figure 1. Schematic of an experimental setup for studying the formation of tangential discontinuities in fluid parameters. 1 - external movable cylinder; 2 - internal stationary cylinder; 3 - investigated liquid; 4 - electric drive of the external cylinder; 5 - light source; 6 - polarizers; 7 - photo or video camera.

Figure 2. External view of the experimental setup.

As a result of the experiments, it was found that glycerol is an optically isotropic liquid medium in the initial state (Figure 3a), which loses its optical homogeneity during the formation of shear deformations (Figure 3b, c). This means that a thin cylindrical layer with a destroyed structure is formed in the medium. It should be emphasized that the time of the destruction of glycerol in the transition layer, as well as the width of the transition layer, depends on the deformation rate. At a constant strain rate, the width of the transition layer with a permitted structure remains constant. After stopping the rotation of the outer cylinder, the structure of glycerol is restored.

3. Conclusion

Experiments have shown that the formation of tangential discontinuity in viscosity is possible in glycerol. Glycerol in the ultra-weak shear range has shear elasticity, therefore, behaves like a non-Newtonian fluid. But at high shear loads exceeding the yield point, the ordered structure is destroyed, glycerin behaves like a Newtonian fluid [13]. Based on the above, when modeling and studying slowly proceeding processes in Newtonian polar fluids, one should take into account their shear elasticity.

a) b) c)

Figure 3. Photos of the formation of tangential discontinuities of shear viscosity in glycerol: a) initial state; b) strain rate 0.5 cm/sec; c) strain rate 5 cm/sec.

The experiment carried out reflects the possibility of the manifestation of momentum transfer mechanisms in liquids associated with the mechanisms of plastic flow and destruction in solids. To confirm this assumption, it is of interest to compare the experimental data on the behavior of fluids under conditions of developed turbulent flow, under conditions of weak shear loads and the widespread case of unstable plastic flow, known as the Portevin-Le Chatelier effect.

These studies can have broad and important areas of application, including the explanation and modeling of the effects of localized (adiabatic) shear under dynamic and shock-wave loading of a wide class of metallic materials. The possible universality of the nature of the development of the instability of condensed media (solid and liquid) under intense influences (in liquids - flows at large Reynolds numbers) is of obvious fundamental and applied interest.

Acknowledgments

This work was carried out with financial support from the Ministry of Science and Higher Education of the Russian Federation (state order No. AAAA-A19-119031890083-9).

References

[1] Frenkel Ya I 1975 *Kinetic Theory of Fluids* (Leningrad: Nauka)
[2] Sakharov A D, Zaidel R M, Mineev V N and Oleynik A G 1964 *Doklady AN SSSR* **159** 1019
[3] Naimark O B 1998 *JETP Lett* **67** 751
[4] Naimark O B 1997 *Pisma Zh. Tekh. Fiz.* **23** 81
[5] Naimark O B 2004 *Advances in Multifield Theories of Continua with Substructure* (Birkhäuser: Bosto) ed. G Capriz and P Mariano p 75
[6] Demetriou M D and Johnson W L 2004 *Journal of applied physics* **95** 2857
[7] Baggioli M, Vasin M, Brazhkin V and Trachenko T 2020 *Physics Reports* **865** 1
[8] Stebenovskij S V 2002 *Zhurnal tekhnicheskoj fiziki* **72** 24
[9] Stebenovskij S V 2004 *Zhurnal tekhnicheskoj fiziki* **74** 21
[10] Glorieux C, Van de Rostyne K, Goossens J, Shkerdin G, Lauriks W and Nelson K A 2006 *Journal of Applied Physics* **99** 013511
[11] Piccirelli R Litovitz T A 1957 *The Journal of the Acoustical Society of America* **29** 1009–20
[12] Rejner M 1965 *Reologiya* (Moscow: Nauka)
[13] Stebenovskij S V 2005 *Prikladnaya mekhanika i tekhnicheskaya fizika* **46** 41
[14] Shramm G 2003 *Osnovy prakticheskoj reologii i reometrii* (Moscow: Koloss)

IOP Conf. Series: Materials Science and Engineering 1093 (2021) 012007 doi:10.1088/1757-899X/1093/1/012007

Numerical study of the stress strain state during tests of polymer composite materials samples

E V Feklistova, A V Babushkin and A V Babushkina

Perm National Research Polytechnic University, Perm, 29, Komsomolsky prospekt, Perm, 614990, Russia

E-mail: cem.feklistova@mail.ru

Abstract. To date, a large number of experimental methods for determining the mechanical properties of composite materials is known. However, the results of these methods. i.e. their comparison features, are relative and give good repeatability in strict compliance with the sample production and experiment technology. In this case, the stress-strain state in the working area of sample structures should be numerically evaluated to determine the generally accepted characteristics. This paper constructs and develops the simplest models of the mechanical behavior of polymer composite material samples. The stress-strain state is numerically simulated in the working areas of samples with various geometric parameters and external loading features. Modeling is made using the ANSYS engineering analysis system. According to the results of calculations for each of the samples, the fields of stresses, deformations, and displacements under compression and tension are obtained for medium models: effective isotropic medium, effective anisotropic medium, and layered composite with effective anisotropic medium texture layers.

1. Introduction

When testing sample structures made of composite materials, non-trivial methods should be used to determine values similar to the basic ones: strength and stiffness. In addition, the standards often recommend using a very rough but simple model to determine these values. In this regard, the issue of evaluating the experimental characteristics of the mechanical properties of sample structures made of composite materials with the parameters of the calculated inhomogeneous deformation and stress fields becomes the most relevant. Moreover, input information for such simulation can be obtained during the basic tests of standard samples, defining the generally accepted characteristics in tension, compression and displacement.

Composite materials have a set of properties and features that differ from traditional structural materials. Hence, they open up wide opportunities for improving existing structures for a wide range of purposes and for developing new structures and technological processes [1].

Three-dimensional reinforcing agents are used for high-performance composite materials in such industries as aerospace, mechanical engineering, aircraft and shipbuilding, metallurgy, nuclear and thermal power industry. Multilayer carbon fiber fabrics are used as a reinforcing material of carbon plastics for difficult and harsh conditions of exposure to high-speed aerodynamic flows, vibration and high temperatures. A review of technological and structural aspects, mechanical behaviors, as well as the current state and prospects for the use of three-dimensional fabrics in polymer composites are given in [2, 3].

A stress riser in a structure leads to a detailed study of the stress and deformable state in the concentration spots under the conditions of elasticity, plasticity and creep. Since the limiting state and destruction occurs precisely in the spots of stress concentration, the assessment of how much a particular raiser will affect the bearing capacity of an element [4-6] is of importance here.

In the numerical study of composite materials, various methods are used, however, the finite element method is very popular [7, 8]. Its main idea is that any continuous quantity, such as temperature, shift or pressure, can be approximated by a discrete model built on a set of piecewise continuous functions defined on a finite number of subdomains. Discrete functions are determined using the values of a continuous quantity at a finite number of points of the area considered. Numerical modeling is used both for assessing the ultimate strength [9] of a structure, and for modeling destruction, using various mechanisms and criteria [10-12].

2. Main part

2.1 Mathematical model

The scientific content of this work is the development of methods for calculating and analyzing the stress-strain state in the working arear of composite materials samples when implementing the composite materials structural-phenomenological approach. The paper considers the numerical FE implementation of a multilevel approach when analyzing the processes of mechanical behavior of composite materials with inhomogeneous components and a complex structure in the working areas of samples structures. Practice has shown that even with the simplest versions of this approach (from considering an isotropic medium to a layered isotropic and layered anisotropic, in this case), it becomes possible to further compare the predicted characteristics, characteristics obtained during basic standard tests, and verify the comparison values, obtained in the course of non-trivial tests of structural samples.

The experimental part of the paper uses a unique scientific installation "Testing and Diagnostic Equipment Set for Studying Structural and Functional Material Properties under Complex Thermomechanical Effects" http://ckp-rf.ru/usu/501309/" at the Center experimental mechanics.

The mathematical model for calculations is the structural and phenomenological model of composite material deformation and fracture [13]. The results of our own experiments and known reference data [14] are used as comparison characteristics. The stresses, in the absence of bulk forces, should satisfy the equilibrium equations:

$$\sigma_{ij,j}(r)=0,\ i,j=1,2,3 \quad (1)$$

Small deformations are associated with displacement by the Cauchy relation

$$\varepsilon_{ij}(r) = \frac{1}{2}(U_{i,j}(r) + U_{j,i}(r)),\ i,j=1,2,3 \quad (2)$$

$$\sigma_{ij} = C_{ijmn}(r)\varepsilon_{mn},\ i,j=1,2,3 \quad (3)$$

Thus, equations 1 to 3 describe the boundary value problem of the mechanics of composites within the framework of the structural and phenomenological model. If the composite properties are considered effective, the system of equations 1-3 is rewritten as:

$$\sigma_{ij,j}^{*}=0,\ i,j=1,2,3 \quad (4)$$

$$\varepsilon_{ij}^{*} = \frac{1}{2}(u_{i,j}^{*}+u_{j,i}^{*}),\ i,j=1,2,3 \quad (5)$$

$$\sigma_{ij} = C_{ijmn}^{*}\varepsilon_{mn}^{*},\ i,j=1,2,3 \quad (6)$$

$$\sigma_{ij}^{*}=\frac{1}{V}\int \sigma_{ij} d^{I}V$$

Equations 4-6 are supplemented with boundary conditions in displacements or stresses on the body surface S:

$$\sigma_{ij}^{*} n_j \Big|_{\Gamma_S} = S_i^o \qquad u_i^0 \Big|_{\Gamma_u} = u_i^o \tag{7}$$

In the general case of the stress-strain state, the generalized Hooke's law for isotropic materials is expressed as:

$$\varepsilon_{11} = \frac{1}{E}[\sigma_{11} - \upsilon(\sigma_{22} + \sigma_{33})]$$
$$\varepsilon_{22} = \frac{1}{E}[\sigma_{22} - \upsilon(\sigma_{11} + \sigma_{33})]$$
$$\varepsilon_{33} = \frac{1}{E}[\sigma_{33} - \upsilon(\sigma_{11} + \sigma_{22})]$$

$\tau_{12} = 2G\varepsilon_{12}, \tau_{23} = 2G\varepsilon_{23}, \tau_{13} = 2G\varepsilon_{13}$

Where the shear modulus is determined through the ratio: $G = \frac{E}{2(1+\upsilon)}$

In the case of an anisotropic material, the components by the tensors of the elastic muzzles C_{ijmn} are determined by the formulas:

$$C_{1111} = \frac{1}{E_{22}A}\left(\frac{1}{E_{33}} - \frac{\upsilon_{23}^2}{E_{22}}\right), \qquad C_{2222} = \frac{1}{E_{22}A}\left(\frac{1}{E_{11}} - \frac{\upsilon_{31}^2}{E_{33}}\right),$$

$$C_{3333} = \frac{1}{E_{11}A}\left(\frac{1}{E_{22}} - \frac{\upsilon_{12}^2}{E_{11}}\right), \qquad C_{1122} = \frac{1}{E_{33}A}\left(\frac{\upsilon_{31}\upsilon_{23}}{E_{22}} + \frac{\upsilon_{12}}{E_{11}}\right),$$

$$C_{1133} = \frac{1}{E_{22}A}\left(\frac{\upsilon_{12}\upsilon_{23}}{E_{11}} + \frac{\upsilon_{31}}{E_{33}}\right); \qquad C_{2233} = \frac{1}{E_{11}A}\left(\frac{\upsilon_{12}\upsilon_{31}}{E_{33}} + \frac{\upsilon_{32}}{E_{22}}\right);$$

$$C_{2323} = G_{23}, \qquad C_{1313} = G_{13}, \qquad C_{1212} = G_{12},$$

$$A = \frac{1}{E_{11}E_{22}E_{33}}\left(1 - 2\,\upsilon_{12}\upsilon_{23}\upsilon_{31} - \frac{E_{11}}{E_{33}}\upsilon_{31}^2 - \frac{E_{22}}{E_{11}}\upsilon_{12}^2 - \frac{E_{33}}{E_{22}}\upsilon_{23}^2\right)$$

Materials composed of alternating flat layers exhibit heterogeneity in a direction perpendicular to the layers. For a composite material with flat layers orthogonal to the x3 axis, it is assumed that perfect contact occurs on all interfaces between structural elements, that is, $[[u_i]] = 0$ and $[[\sigma_{i3}]] = 0$. Then, the closed system of equations includes equilibrium equations, which have only one summand, where differentiation along the x_3 axis occurs:

$$\sigma_{i3,3} = 0 \tag{8}$$

The geometric relationships that establish the relationship between structural deformations and displacements are as follows:

$$\varepsilon_{ij} = \langle\varepsilon_{ij}\rangle + \frac{1}{2}\left(u_{i,3}\delta_{i3} + u_{j,3}\delta_{j3}\right) \tag{9}$$

Then the constitutive relations take the form:

$$\sigma_{ij} = C_{ijmn}\varepsilon_{mn} \tag{10}$$

Boundary conditions

$$\langle\varepsilon_{ij}\rangle x_j \Big|_{\Sigma} = u_i^o \qquad \langle\sigma_{ij}\rangle n_j \Big|_{\Sigma} = S_i^o \tag{11}$$

are equivalent to specifying macrodeformations $\varepsilon_{ij}^{*} = \langle\varepsilon_{ij}\rangle$ or macrostresses $\sigma_{ij}^{*} = \langle\sigma_{ij}\rangle$.

2.2 Results of numerical calculations

To determine the fields of stresses, deformations and displacements under compression and tension, the deformation of structural elements, differing in geometric parameters and external loading features, is numerically simulated. Numerical modeling is carried out using the ANSYS engineering analysis system.

Three types of tests are simulated: compression of sample strips, compression of holed samples, bending of a curved beam. The geometry and dimensions of the samples comply with test standards ASTM D3039, ASTM D3410, ASTM D5766, ASTM D6484 and ASTM D641. We use the results of our own research as the isotropic material properties, namely, data from tests — Young's modulus and Poisson's ratio [14]; data for calculating an anisotropic material are taken from a reference book. The isotropic material properties are given in table 1 and the anisotropic material properties are presented in table 2.

Table 1. Properties of isotropic material.

E, Pa	υ
66	0,3

Table 2. Properties of anisotropic material.

E_x, Pa	E_y, Pa	E_z, Pa	υ_{xy}	υ_{zx}	υ_{xy}	G_{yz}, Pa	G_{zx}, Pa	G_{xy}, Pa
$58\cdot10^9$	$29\cdot10^9$	$6\cdot10^9$	0.2	0.2	0.2	$3\cdot10^9$	$2\cdot10^9$	$4\cdot10^9$

For each of the samples, grid models are built (Figure 1), using a 20-node finite element SOLID 186, a three-dimensional structural solid element with 20 nodes. This type of element has a quadratic displacement behavior and is well suited for modeling both regular and irregular grids. The accuracy of the numerical study is achieved through a preliminary analysis for the convergence of the solution when the grid is clustered. The parameters of the grid model in terms of the number of nodes and the quality of the grid model are given in Table 3. The improved grid quality parameters of the sample models considered are justified by simplified geometry. The grid model is adopted near the stress concentrators (sample hole and bend). When carrying out the calculations, a structured grid is used. It has not only a more advantageous application in terms of efficient use of computing resources, but also the ability to quickly change the parameters of the grid model. It allows to quickly and conveniently control the grid creation, which is important especially in areas that have characteristic geometric features.

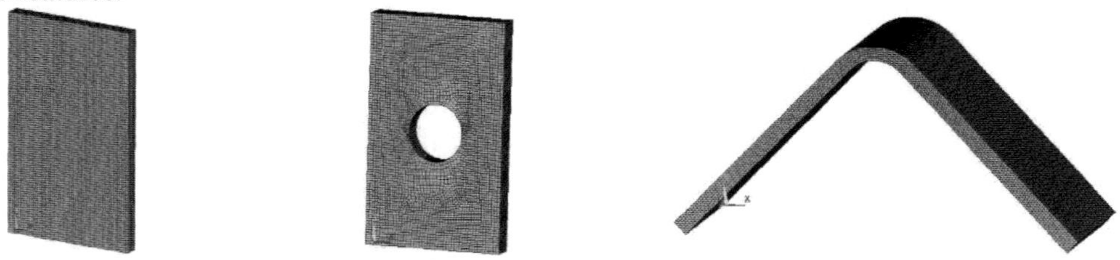

a b c

Figure 1. Mesh models of the sample - strip (a), sample with a hole (b), curved beam (c).

Table 3. Parameters of mesh models.

Parameter	Mesh structure model	Mesh model of a sample with a hole	Curved Beam Mesh Model
Number of nodes	143022	139592	297077
Amount of elements	31122	30944	64872

When numerically modeling, the second stage after the grid is built should set the boundary conditions (Figure 2). For samples with or without a hole for the bottom surface, the boundary condition for limiting the displacement along three coordinates is set, while, for the opposite surface, the condition of tension or compression by 100 Pa is set. For a curved beam sample, the lower planes of the surfaces has a displacement restriction in two coordinates; and displacement is allowed along the third coordinate. A load of 100 Pa is applied to the upper part of the sample for compression or tension.

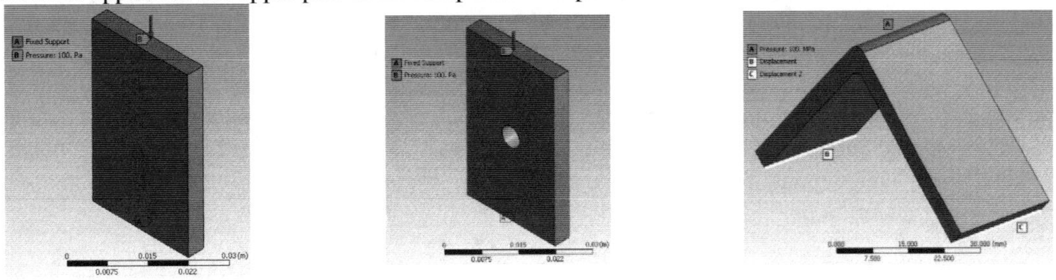

Figure 2. Boundary conditions of sample - strip (a), sample with hole (b), curved beam (c).

For each of the above plates, as well as the curved beam, a calculation is made under compression and tension by 100 Pa. According to the results of calculations for each of the samples, the fields of stresses, strains and displacements under compression and tension are given for the medium models: effective isotropic medium, effective anisotropic medium, and layered composite with effective anisotropic medium texture layers (Figure 3).

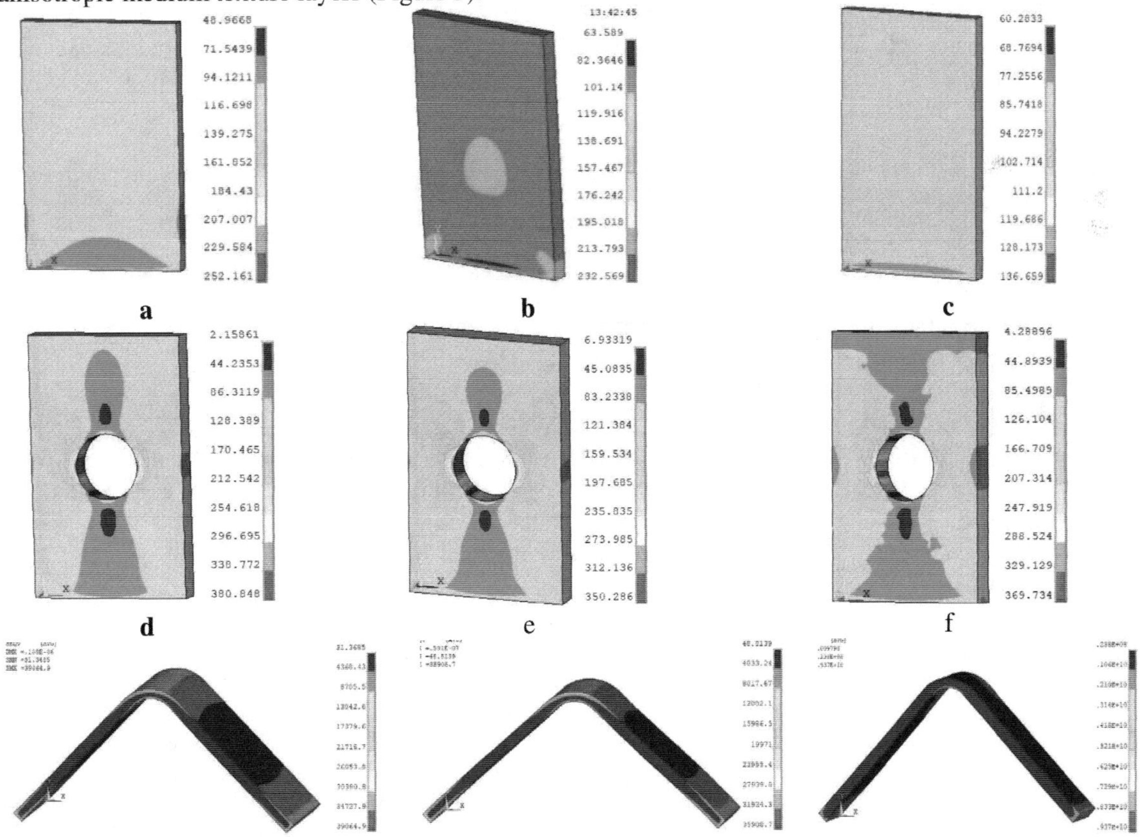

Figure 3. Fields of distribution of equivalent stresses in compression of strip specimens (a-c), compression of specimens with a hole (d-f), curved beam (g-i) for models of the medium: effective isotropic medium (a, d, g), effective anisotropic medium (b, e, h) and a layered composite with texture layers of an effective anisotropic medium (c, f, i).

3. Conclusion

Thus, this paper constructs a boundary value problem of deformation and fracture of composite materials to determine the complex stress-strain state in the working areas of samples of medium models: effective isotropic medium, effective anisotropic medium, and layered composite with effective anisotropic medium texture layers. To determine the fields of stresses, deformations and displacements under compression and tension, the deformation of sample structures is numerically simulated. Numerical modeling is performed using the ANSYS engineering analysis system. Three types of tests are simulated: compression of sample strips, compression of holed samples, and bending of a curved beam. Sample geometry and dimensions comply with ASTM test standards. According to the results of calculations for each of the samples, the fields of stresses, deformations and displacements under compression are obtained.

Acknowledgements

This work was carried out in Perm National Research Polytechnic University using the equipment of the Centre of Collective Usage «Center of experimental mechanics» http://www.ckp-rf.ru/ckp/353547/ with the use of the results obtained in the study supported by The Russian Foundation for Basic Research (project No. 18-01-00763).

References

[1] Lubin J 1988 Reference book on composite materials (Moscow: Mechanical Engineering) 584
[2] Sharma P, Mali H S, Dixit A Mechanical behavior and fracture toughness characterization of high strength fiber reinforced polymer textile composites *Iranian Polymer Journal* 2020
[3] Asenjan M S, Sabet S A R, Nekoomanesh M 2020 *Iran Polym J* **29** 301–307
[4] Senthil K, Arockiarajan A, Palaninathan R, Santhosh B, Usha K M 2013 *Compos Struct* **106** 139-149
[5] Bespalov V A, Gotselyuk T B, Kovalenko N A, Olegin I P 2015 *Omsk Scientific Bulletin* **3 (143)** 329-333
[6] Ataş A, Soutis C 2013 *Compos. Part B* **54** 20–27
[7] Zenkevich O , Chang I 1967 The finite element method in the theory of structures and in mechanics of continuous media (New York) 240
[8] Segerlind L 1979 Application of the finite element method (Moscow: MIR) 392
[9] Aidi B, Case S W 2015 *Compos Mater* **22** 837–855
[10] Bian T, Guan Z, Liu F 2018 *Archive of applied mechanics* 1-20
[11] Guo Z, Zhu H, Li Y, Han X, Wang Z 2016 Compos. Mater **23** 12091218
[12] Kumar D, Roy R, Kweon J-H, Choi J-H 2016 Compos. Mater. **23** 397419
[13] Wildeman V, Sokolkin Yu, Tashkinov A 1998 The mechanics of inelastic deformation and fracture of composite materials (Moscow: Nauka) 288
[14] Babushkin A V , Babushkina A V, Strungar E M, Staroverov O A, Lobanov D S, Temerova M S, Feklistova E V 2019 Procedia Structural Integrity **17** 658-665

Dielectric properties of solid solutions in a PFN-based system

E V Glazunova, I A Verbenko, I A Shilkina, A V Nagaenko and L A Reznichenko

Institute of Physics, Southern Federal University, 194, Stachki Ave., Rostov-on-Don, 244090, Russia

E-mail: kate93g@mail.ru

Abstract. The results of X-ray diffraction analysis of the solid solutions in the (1-x)Pb($Fe_{1/2}Nb_{1/2}$)O_3–xPb($Fe_{2/3}W_{1/3}$)O_3 systems with x = 0.0÷0.5, investigating their microstructure and measurements of dielectric characteristics are presented in this paper. The critical influence of Pb($Fe_{2/3}W_{1/3}$)O_3 on the crystallization of solid solutions and the nature of grain boundaries is established. A sintering mechanism of the solid solutions is defined. Correlations between the dynamics of the crystalline and grain structures and the dielectric properties of investigating solid solutions are established.

1. Introduction

In recent years, special attention in the field of materials science has been paid to materials with a complex ordering of dipole and magnetic moments (multiferroics). One of the most promising multiferroics is lead ferroniobate Pb($Fe_{1/2}Nb_{1/2}$)O_3 (PFN). PFN has two ferroelectric phase transitions at temperatures Tc ≈ 380 K and Tc ≈ 350 K [1, 2] and two magnetic phase transitions (PT) at temperatures T_{N1} ≈150K and T_{N2} ≈10 K [3, 4]. It is also known that the compounds having a perovskite-like structure type (AB'B''O_3) can be completely or partially disordered in the B-position [5-6]. It contributes to the emergence of the unusual physical properties in such materials. The most promising of them can be used in such modern fields as spintronics and straintronics [7-9].

Thus, investigations aimed to establish the patterns of structure formation and formation of dielectric responses in solid solutions based on the high-temperature multiferroic Pb($Fe_{1/2}Nb_{1/2}$)O_3 are relevant.

2. Experimental section

Systems (1-x)Pb($Fe_{1/2}Nb_{1/2}$)O_3—xPb($Fe_{2/3}W_{1/3}$)O_3 with $0.0 \leq x \leq 0.5$ were the subjects of this study. The solid solutions were prepared using a two-stage solid-state reaction at T_1=1073 K for 4 h and T_2=(1123-1173) K for 4-10 h followed by sintering using conventional ceramic technology at $T_{sint.}$=(1143÷1373) K for 2 h. The raw powders were PbO (98%), Fe_2O_3 (99%), WO_3 (99.9%), Nb_2O_5 (98%).

X-ray diffraction was conducted using a DRON 3.0 diffractometer (using CoK_α radiation with Bragg-Brentano focusing). The content of impurity phases was estimated from the relative intensity of their strong line: $I/I_I \cdot 100\%$, where I is the intensity of the strong line of the impurity phase, I_I is the intensity of the strong line of the perovskite phase.

The microstructural measurements were performed by using a JSM – 6390L scanning electron microscope.

The dependencies of the real ε′ and imaginary ε″ parts of dielectric permittivity were measured on the mechanically free samples ∅10x1 mm in a temperature range T = 320-600 K at frequencies of 20

Hz-1 MHz using Agilent 4980A LCR meters and Wayne Kerr 6500B precision LCR meters. The samples had annealed for 6 hours before the measurements. The measuring step was 2K.

3. Results and discussion

An X-ray phase analysis of the $(1\text{-}x)Pb(Fe_{1/2}Nb_{1/2})O_3$—$xPb(Fe_{2/3}W_{1/3})O_3$ system with $0.0 \leq x \leq 0.5$ showed that the solid solutions had not impurity phases. All of the samples have a cubic structure at room temperature (Figure 1a).

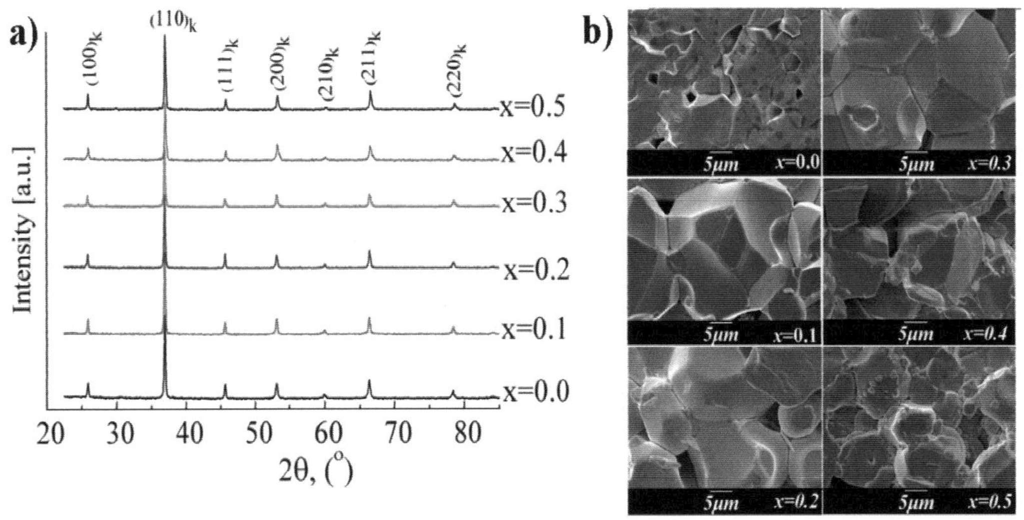

Figure 1. X-ray diffraction peaks (a); fragments of the microstructure (b) of the solid solutions in the $(1\text{-}x)Pb(Fe_{1/2}Nb_{1/2})O_3$—$xPb(Fe_{2/3}W_{1/3})O_3$ system.

A microstructural analysis demonstrated that when a small amount of $Pb(Fe_{2/3}W_{1/3})O_3$ ($x = 0.1$) are added to the system the grain size is significantly increased (d = 25 μm). After reaching a concentration of x=0.2 a liquid phase is formed. This is reflected in the appearance of the double boundaries of grains, intercrystalline interlayers that form continuous channels of the liquid phase. In doing so the habit of the grains takes the round, spherical shape ($x = 0.2$, $x = 0.3$). A further increase in the concentration of $Pb(Fe_{2/3}W_{1/3})O_3$ results in the growing number of the liquid phase. During the process, the smaller particles are dissolved and deposited on the boundaries of larger grains ($x = 0.5$). Further, the intercrystalline interlayers completely are filled with a liquid phase and the grain boundaries become indistinguishable. Thus, it can be concluded that sintering takes place by the mechanism of surface diffusion.

The temperature dependencies of the real (ε') and imaginary (ε'') parts of the dielectric constant are shown in Figure 2. The dielectric constant ε' is characterized by strong dispersion before and at the moment of the phase transition.

The analyses of the dependencies demonstrated that ε' and ε'' pass through the maximum blurred by temperature. When x=0.1 the dependencies of ε' become more pronounced. However, with increasing concentration of the $Pb(Fe_{2/3}W_{1/3})O_3$ the values of the dielectric constant in the maximum are decreased and shifted to lower temperatures.

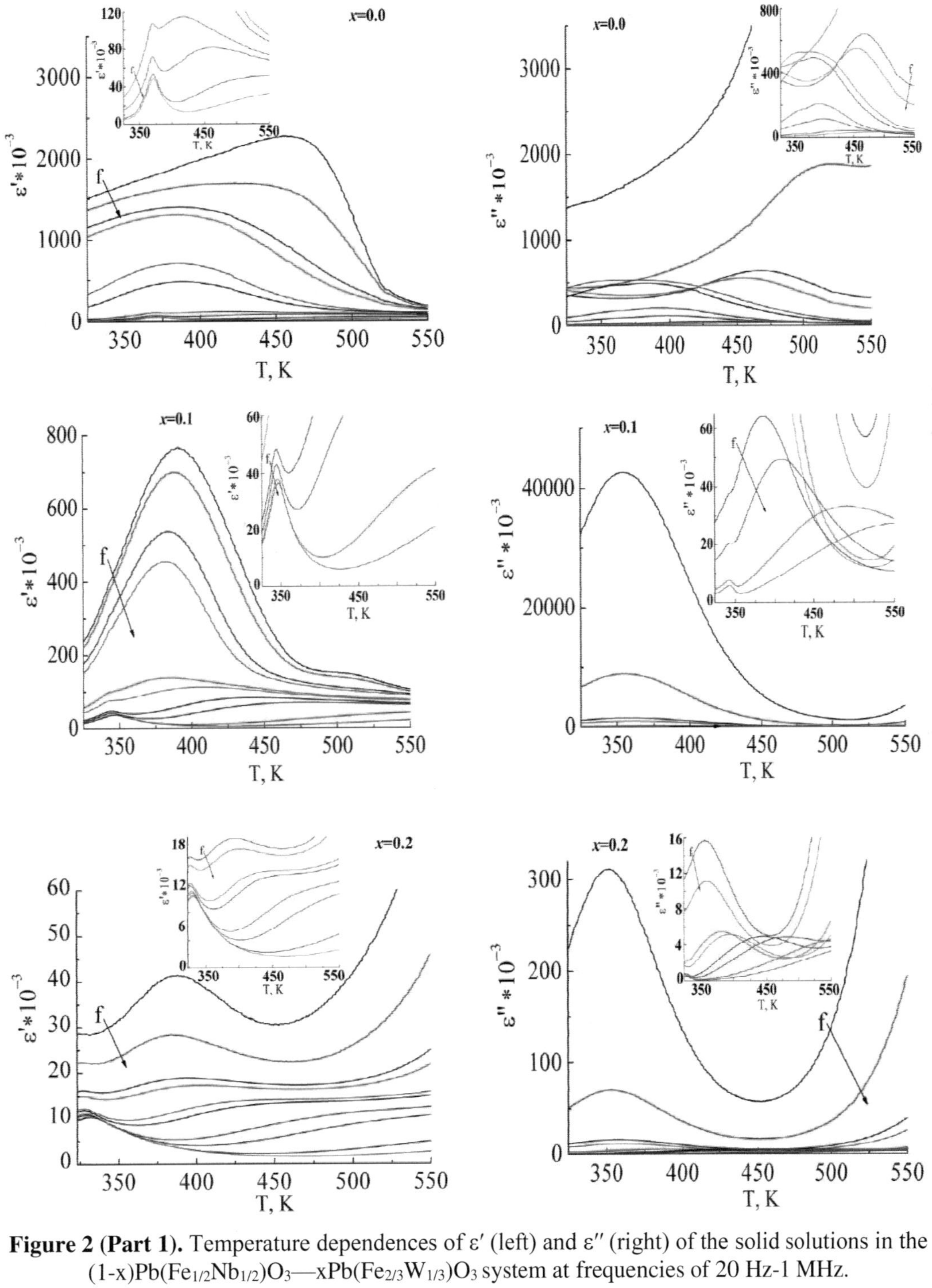

Figure 2 (Part 1). Temperature dependences of ε′ (left) and ε″ (right) of the solid solutions in the $(1\text{-}x)Pb(Fe_{1/2}Nb_{1/2})O_3$—$xPb(Fe_{2/3}W_{1/3})O_3$ system at frequencies of 20 Hz-1 MHz.

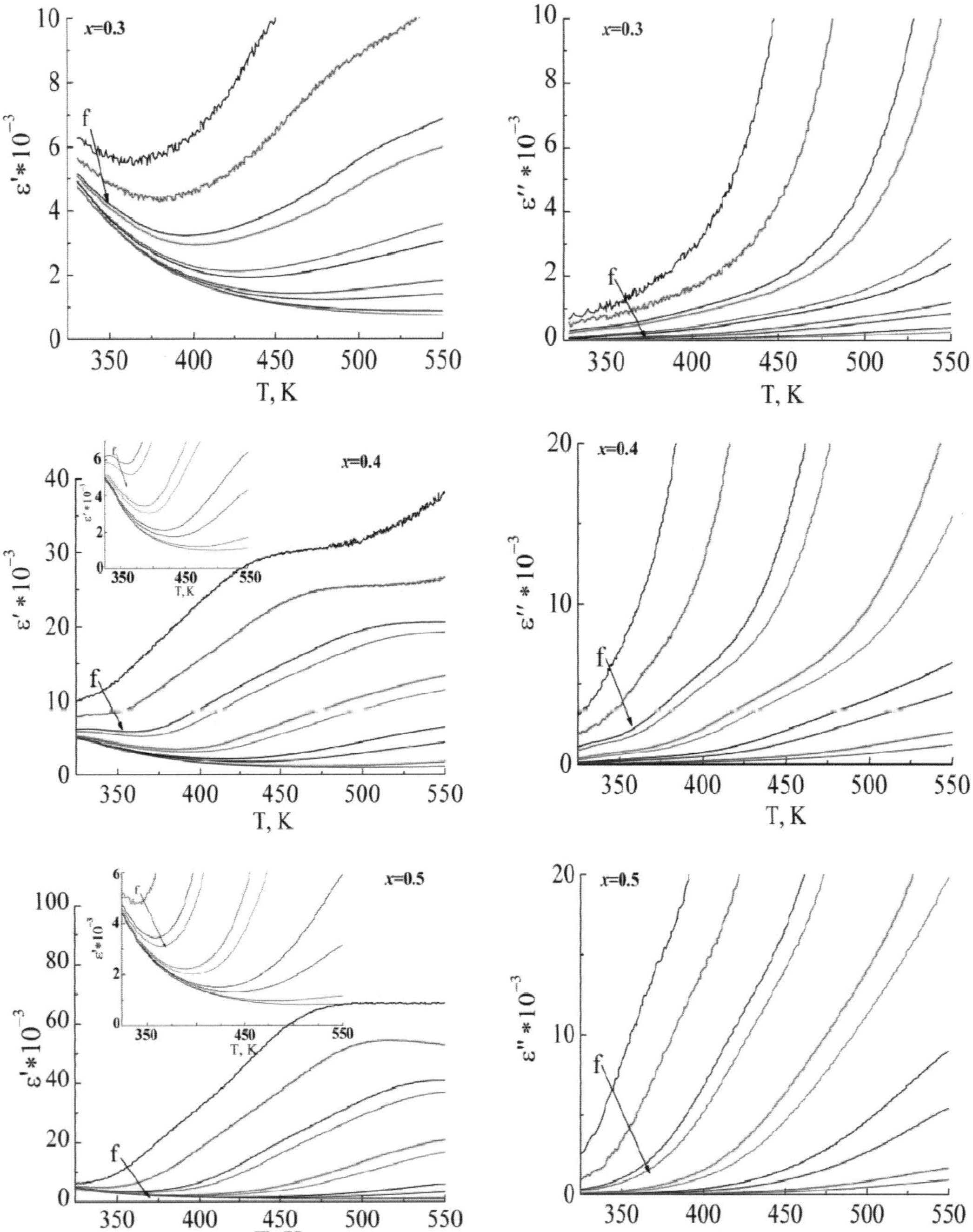

Figure 2 (Part 2). Temperature dependences of ε' (left) and ε'' (right) of the solid solutions in the $(1\text{-}x)Pb(Fe_{1/2}Nb_{1/2})O_3$—$xPb(Fe_{2/3}W_{1/3})O_3$ system at frequencies of 20 Hz-1 MHz.

The ε′ passes through a blurred maximum which shifted to a higher temperature at increasing frequency. However, the imaginary part is tens of times higher than the real part at low frequencies. Comparison of these values leads to the conclusion that such behavior of the temperature dependencies of the ε ' and ε" is associated with high electrical conductivity of the solid solutions [10]. The nature of the electrical conductivity dependence on the temperature and frequency of the field may be due to the movement of defects due to the restructuring of the structure or the movement of vacancies resulting from the reduction of variable valence ions (Fe, Nb).

4. Conclusions

The correlations between phase formation, dynamics of the microstructure, and dielectric responses of the solid solutions in the (1-x)$Pb(Fe_{1/2}Nb_{1/2})O_3$—x$Pb(Fe_{2/3}W_{1/3})O_3$ system were established. The critical impact of a liquid phase on crystallization and macroscopic responses in the solid solutions investigated was found.

Acknowledgments
The reported study was funded by RFBR, project number 19-32-90099\19 using the equipment of the Collective Use Center "Electromagnetic, Electromechanical and Thermal Properties of Solids" of the Research Institute of Physics and President's scholarship 2175.2021.1.

References

[1] Lampis N, Sciau P, Lehmann A G, 1999 Rietveld refinements of the paraelectric and ferroelectric structures of $PbFe_{0.5}Nb_{0.5}O_3$ *J Phys Condens.: Matter* **11** 17 pp 3489–5000
[2] Korotkov L N *et al* 2009 Structure and electrophysical properties of amorphous material based on lead ferronobat *J Tech Phys.* **79** 8 pp 62-70
[3] Kleemann W, Shvartsman V, Borisov P, Kania A 2010 Multiferroic and magnetoelectric materials -Developments and perspectives *Phys. Rev. Lett* **105** pp 202-257
[4] Gruszka I, *et al* 2018 Characterization of multiferroic $PbFe_{0.5}Nb_{0.5}O_3$ and $PbFe_{0.5}Ta_{0.5}O_3$ ceramics derived from citrate polymeric Precursors *J. American Ceramic Society* pp 15998 (DOI:10.1111/jace.15998)
[5] Reznichenko L A, *et al* 2011 Ferroelectrics, relaxors, multiferroics with different thermodynamic prehistory *Nova Science Publishers* (Monograph «PiezoceramicMaterials and Devices) **4** pp 1-44
[6] Raevski I P, *et al* 2002 Study of Cation Ordering and Magnetic Phase Transitions in Ternary Fe-Containing Perovskite Oxides by Mössbauer Spectroscopy *Crystallography Reports* **47** 6 pp 1012–1015
[7] Ramesh R. and Spaldin N A 2007 Multiferroics: progress and prospects in thin film *Nature Materials* **6** pp 21
[8] Kimura Koji *et al* 2019 Local structural analysis of $Pb(Fe_{1/2}Nb_{1/2})O_3$ multiferroic material using X-ray fluorescence holography *Japanese Journal of Applied Physics* **58** pp 100601
[9] Bochenek D, *et al* 2011 Multiferroic ceramics $Pb(Fe_{1/2}Nb_{1/2})O_3$ doped by Li *J. Electroceram* **26** pp 8-13
[10] Vezzoli G G 1982 Electrical properties of NbO_2 and Nb_2O_5 at elevated temperature in air and flowing argon *Phys.Rev. B.* **26** 7 pp 3954-3957

IOP Conf. Series: Materials Science and Engineering 1093 (2021) 012009 doi:10.1088/1757-899X/1093/1/012009

Phase formation, structure, and properties of the solid solutions in a system based on $K_{0.5}Na_{0.5}NbO_3$ and $Ca_2Nb_2O_7$

E V Glazunova, I A Verbenko, I A Shilkina, A V Nagaenko and L A Reznichenko

Institute of Physics, Southern Federal University, 194, Stachki Ave., Rostov-on-Don, 244090, Russia

E-mail: kate93g@mail.ru

Abstract. This study reports the creation of the solid solutions $(1\text{-}x\text{-}y)KNbO_3 - yNaNbO_3 - xCa_2Nb_2O_7$ system, with y=0.50, $0.0 \leq x \leq 0.1$, Δx=0.025 using solid-state reaction and sintering by conventional ceramic technology. The mechanisms and patterns of formation of the microstructure and dielectric properties of received solid solutions are established. The most probable mechanism of the formation of defects in the structure of $(1\text{-}x\text{-}y)KNbO_3 - yNaNbO_3 - xCa_2Nb_2O_7$ solid solutions was determined.

1. Introduction

Knowing the importance of preserving the environment resulted in the adoption of a directive banning the use of toxic elements (RoHS 2002/95/EC) [1] in some fields of science and technique. In this regard, lead-free piezoceramic has become increasingly popular among researchers in recent years. One of the most promising bases for lead-free piezo materials is the system of potassium-sodium niobate (K, Na)NbO_3 (KNN). In recent decades, this system has been intensively investigating due to the proximity of its piezoelectric characteristics to the $PbZr_{1-x}Ti_xO_3$ (PZT) materials used in industry [2] as well as through the possibility to produce materials with a large amount of stored energy [3]. However, producing KNN-ceramic is difficult due to the properties of materials are dependent on thermodynamic prehistory (conditions of the production) [4-6]. It is known from the literature that obtaining ceramic based on KNN-system is associated with several chemical and technological difficulties. Such as hydrolysis of parent substances, a narrow spacing optimal synthesis and sintering temperatures, high volatility of alkaline components at relatively low temperatures that lead to producing ceramic with low density and high conductivity. These problems are being addressed through the introduction of various modifiers.

Pyroniobate calcium ($Ca_2Nb_2O_7$) has a high temperature of phase transition~ 1700 K [7]. Solid solutions of the KNN-based systems have high values piezoelectric characteristics. Consequently, in conjunction, they can provide the right combination of properties to develop modern generation high-temperature devices.

Thus, the establishing of the correlations between phase formation, structure, and properties in the $(1\text{-}x\text{-}y)KNbO_3 - yNaNbO_3 - xCa_2Nb_2O_7$ solid solutions are relevant.

2. Experimental section

The $(1\text{-}x\text{-}y)KNbO_3 - yNaNbO_3 - xCa_2Nb_2O_7$ solid solutions with y=0.50, $0.0 \leq x \leq 0.1$, Δx=0.025 were the subjects of this study. The solid solutions were prepared using a two-stage solid-state reaction at T_1=1250-1270 K, T_2=1400-1420 K for 6 h followed by sintering using conventional ceramic technology at $T_{sint.}$ = 1400-1550 K for 2 h. The raw powders were $NaHCO_3$ (99 %), $KHCO_3$ (99 %), Nb_2O_5 (98 %), and pre-synthesized $Ca_2Nb_2O_7$.

X-ray diffraction was conducted using a DRON 3.0 diffractometer (using CoK_α radiation with Bragg-Brentano focusing). The content of impurity phases was estimated from the relative intensity of their strong line: $I/I_1 \cdot 100\%$, where I is the intensity of the strong line of the impurity phase, I_1 is the intensity of the strong line of the perovskite phase.

The microstructural measurements were performed by using a JSM – 6390L scanning electron microscope.

The dielectric permittivity was measured in a temperature range T = 300-900 K at frequencies of 1 kHz-1 MHz using Agilent 4980A and Wayne Kerr 6500B precision LCR meter.

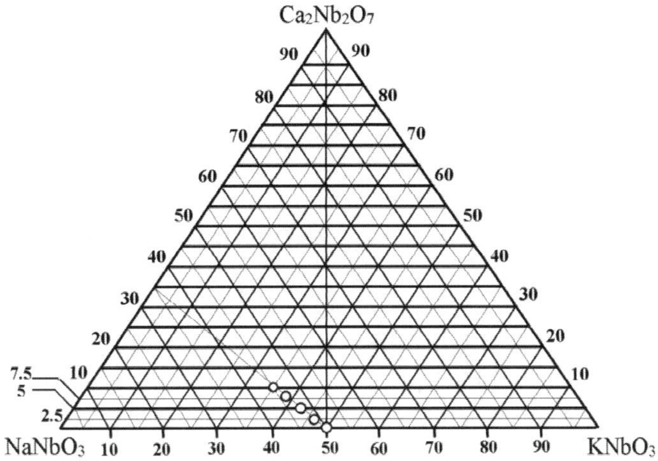

Figure 1. Triangle of Gibbs phases of the (1-x-y)$KNbO_3$ – $y$$NaNbO_3$ – $x$$Ca_2Nb_2O_7$ system with marked experimental points.

3. Results and discussion

An X-ray phase analysis of the (1-x-y)$KNbO_3$ – $y$$NaNbO_3$ – $x$$Ca_2Nb_2O_7$, with y=0.50, 0.0≤x≤0.1, Δx=0.025 system showed that all of the solid solutions had perovskite-type structures.

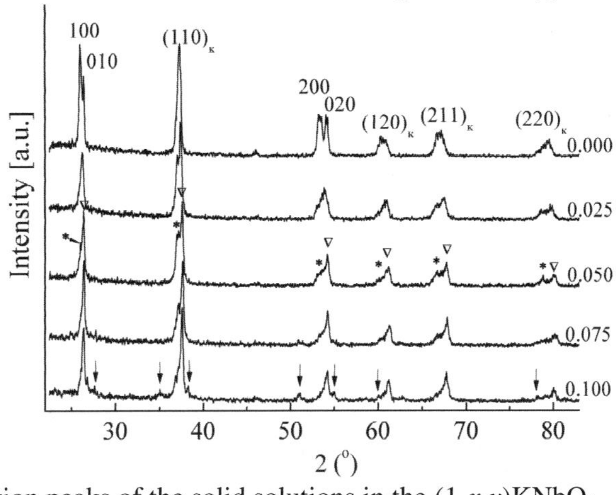

Figure 2. X-ray diffraction peaks of the solid solutions in the (1-x-y)$KNbO_3$ – $y$$NaNbO_3$ – $x$$Ca_2Nb_2O_7$ system (the pseudocubic (PsC) phase peaks are marked by stars, the cubic (C) phase peaks are marked by triangles, and the layered perovskite-like phase peaks are marked by arrows).

In figure 2 is shown that on X-ray patterns of the solid solutions with x=0.00-0.05 the indices of the diffraction peaks are corresponding to the perovskite axes. The X-ray phase analysis also demonstrated that the solid solutions with x=0.10 [8] had peaks of the layered perovskite phase. Traces of this phase are also visible on the X-ray diffraction pattern of the solid solution with x = 0.075.

The initial solid solution $K_{0.5}Na_{0.5}NbO_3$ has a monoclinic distorted subcell. When the $Ca_2Nb_2O_7$ is introduced into the parent solid solution, it decays into two solid solutions with larger and smaller cell parameters. The symmetry of a solid solution with a smaller cell parameter is cubic. The solid solution with a large parameter is very heterogeneous, therefore, it is not possible to reliably determine its symmetry. We designated its symmetry as pseudocubic. When increased the concentration of $Ca_2Nb_2O_7$ the volume of the cubic cell decreases that associated with K→Ca substitution (R (K^+)=1.33 Å R (Ca^{2+})=1.04 Å) (see Table 1).

Table 1. Parameters of cell and density of ceramics of the (1-x-y)$KNbO_3$ – $y$$NaNbO_3$ – $x$$Ca_2Nb_2O_7$ solid solutions

x	symmetry	a, Å	b, Å	β (°)	V, Å^3	P_{exp} g/cm^3	P_{xr} g/cm^3	P_{rel}, %
0.000	M	3.997	3.938	90.29	62.91	3.89	4.54	85.73
0.025	M+	3.990	3.935	90.31	62.63	3.95	4.64	85.19
	PsC	4.005			64.26			
0.050	C+	3.937			61.01	3.68	4.89	75.25
	PsC	3.990			63.53			
0.075	C+	3.934			60.91	3.66	5.03	72.77
	PsC	3.981			63.38			
0.100	C	3.936			60.97	3.80	5.22	72.79

As the concentration of $Ca_2Nb_2O_7$ increases in the range of $0.00 \leq x \leq 0.10$, a number of phase transitions occur in this system: monoclinic (M) → M+ pseudocubic (PsC) → PsC +cubic (C) (see figure 3 and Table 1).

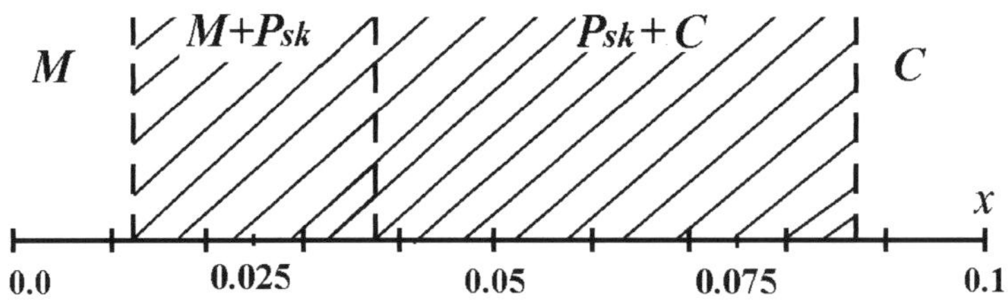

Figure 3. Phase diagram (1-x-y) $KNbO_3$ – $y$$NaNbO_3$ – $x$$Ca_2Nb_2O_7$ system.

Fragments of microstructures of the (1-x-y)$KNbO_3$ – $y$$NaNbO_3$ – $x$$Ca_2Nb_2O_7$ solid solutions are shown in figure 4. The analysis of figure 4 shows that pure $Na_{0.5}K_{0.5}NbO_3$ has a cubic form grain. The grain boundaries are equilibrium. The packing of the structure is dense and consists of large crystallites with an average grain size of d ≈ 15 μm. If the concentration of the component x increases, the grain size decreases significantly, but the character of the grain landscape remains up to x = 0.05. An anisometric type of grain structure is observed in which needle and lamellar grains which are characteristic to layered pyroniobates, which can be nuclei of a new layered phase [9] are formed on the background of cubic crystallites (x above 0.075). The orintation of such grains is chaotic therefore different sections of elongated prisms are observed. The size of such grains is 5-10 μm.

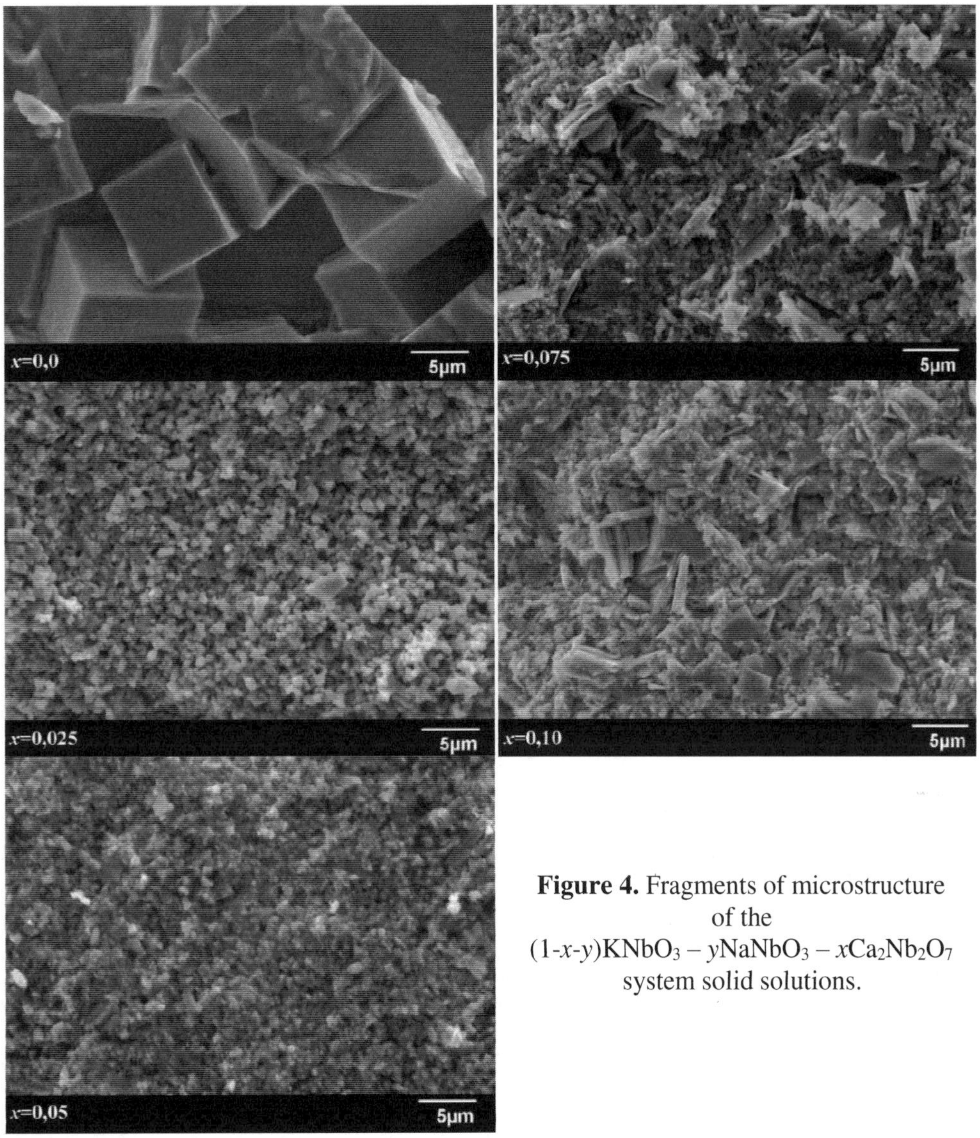

Figure 4. Fragments of microstructure of the $(1\text{-}x\text{-}y)KNbO_3 - yNaNbO_3 - xCa_2Nb_2O_7$ system solid solutions.

The temperature dependence of dielectric constant demonstrated two extrema in the (480–550) K and (680-750) K regions at x=0.0, corresponding to phase transitions between the rhombohedral→tetragonal and tetragonal→cubic phases, respectively. The dependences of the $\varepsilon/\varepsilon_0$ at x = 0.025 and x = 0.05 are characterized by a strong frequency dispersion (starting at 550 K) at the point of the phase transition at a frequency of 1 kHz. The maxima of the dielectric constant on the dependences are weakly. At frequencies above 1 kHz, the strong dispersion begins to appear at 700K, and shifted as the frequency increases to the higher-temperature region (figure 5).

This behavior may be related to the increasing number of free charge carriers at domain boundaries emerging from the adding of bivalent calcium into the system. The charged boundaries of grains interact with an alternating electric field, which leads to the appearance of Maxwell-Wagner type polarization. Further, contribution to the polarization mechanism may be due to the appearance of ceramic-air interphases during the formation of the pores and vacancies. Let's consider the possible means of their formation. At producing solid solutions with a perovskite-type structure, the adding of compounds like $Ca_2Nb_2O_7$ leads to an excess of oxygen octahedrons concerning the cations occupying the A-position.

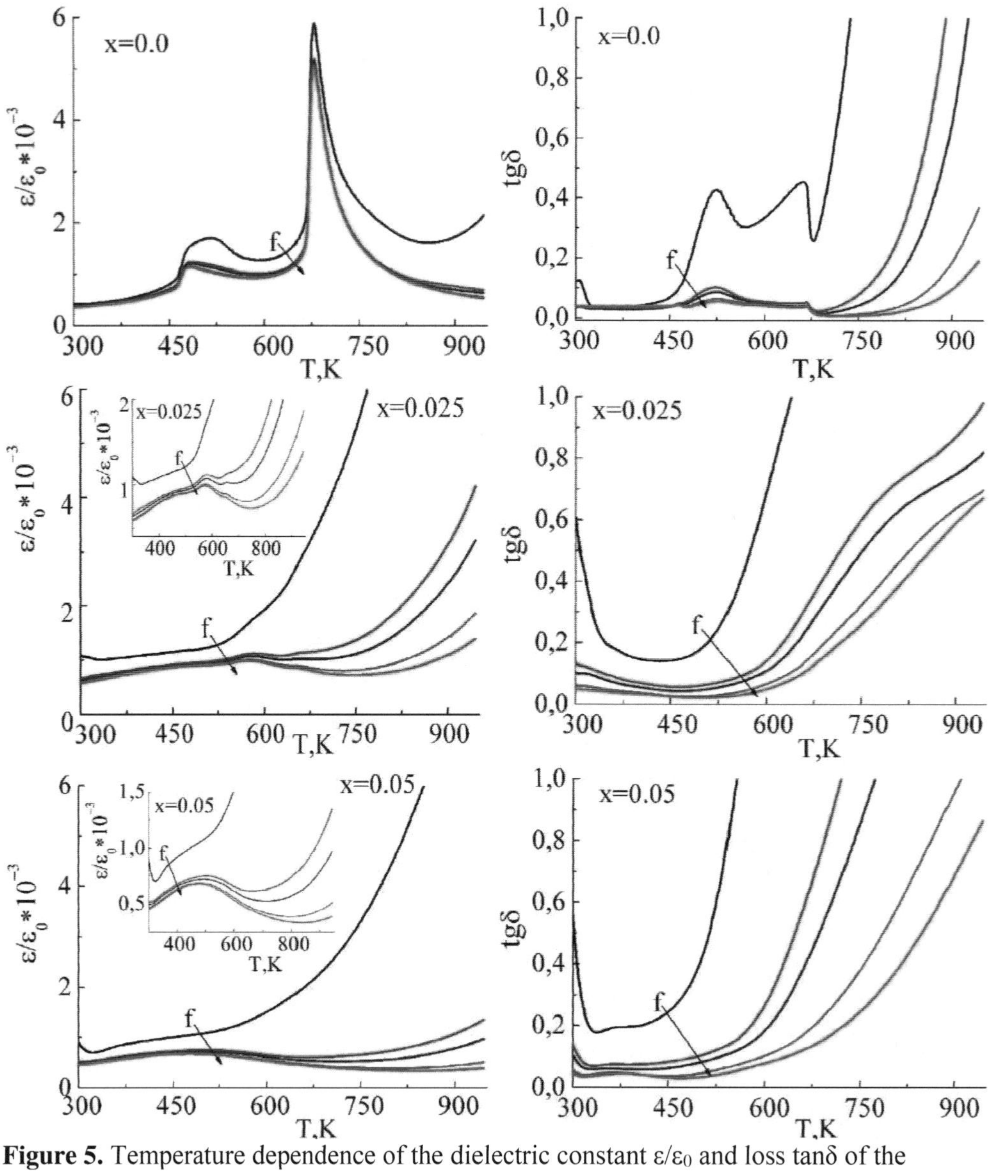

Figure 5. Temperature dependence of the dielectric constant $\varepsilon/\varepsilon_0$ and loss tanδ of the (1-*x*-*y*)$KNbO_3$ – *y*$NaNbO_3$ – *x*$Ca_2Nb_2O_7$ solid solutions with a) *x*=0.0; b) *x*=0.025; c) *x*=0.05 at frequencies of 1 kHz-1 MHz.

At low concentrations $Ca_2Nb_2O_7$, such an excess can be offset by the formation of vacancies in the A sublattice, which must be reflected in increasing the unit cell volume of the perovskite structure $A^{1+}_{1-x}M^{2+}_x(Nb^{5+}O^{2-}{}_3)_{1+x}$, where M^{2+}_x – divalent cation.

However, this trend is less pronounced when compared to potassium replacement by the smaller calcium Ca (R (Ca2+) =1.04 Å, R (K+) =1.33 Å). Furthermore, the clustering and annihilation of vacancies take place when their accumulation in the A-sublattice via the crystallographic shift with the formation of linear, planar, and block superstructures [10].

Furthermore, the partial loss of oxygen with the formation of vacancies occurs at temperature rise, as in any oxide compounds. Besides, the relatively low reduction temperature of niobium-containing complex oxides (≈450 ° C) deals with a variable valence of Nb and is correlated with increases in electrical conductivity at increasing frequency in the compounds of this group [11]. This process lasts up to 750 K. Reduction of niobium occurs according to the scheme $A^{1+}Nb^{5+}_{1-x}Nb^{4+}_{x}O^{2-}_{3-x/2}vac_{x/2}$, where $vac_{x/2}$ - vacancies in the oxygen sublattice, with the emergence of point defects. The interaction of two types of coexisting defective systems can lead to additional Maxwell-Wagner type maximums, changes in the nature of dispersion processes and their energy characteristics.

Conclusions

The correlations between composition, crystal structure, microstructure, and dielectric responses in the $(1\text{-}x\text{-}y)KNbO_3 - yNaNbO_3 - xCa_2Nb_2O_7$ solid solutions were established. The most likely mechanism of the formation of defects in the structure was established.

Acknowledgments

The study was carried out with the financial support of the Ministry of Science and Higher Education of the Russian Federation (State task in the field of scientific activity, scientific project No (0852-2020-0032)/(BA30110/20-3-07IF) using the equipment of the Collective Use Center "Electromagnetic, Electromechanical and Thermal Properties of Solids" of the Research Institute of Physics.

References

[1] Directive 2002/95/EC_of the European Parliament and of the Council of 27 January 2003 on the restriction of the use of certain hazardous substances in electrical and electronic equipment *Official Journal of the European Union* **13** pp 10–24

[2] Lee M K , Yang S A, Park J J and Lee G J 2013 Proposal of a rhombohedral-tetragonal phase composition for maximizing piezoelectricity of $(K,Na)NbO_3$ ceramics *Scientific Reports* **9** pp 4195

[3] Xiang Zh, Di H, Zhongbin P *et al* 2021 Enhancement of recoverable energy density and efficiency of lead-free relaxor-ferroelectric BNT-based ceramics *Chemical Engineering Journal* **406** pp 126818

[4] Reznitchenko L A *et al* 2012 Ferroelectrics, relaxors, multiferroics with different thermodynamic prehistory *Nova Science Publishers* (Monograph «Piezoceramic Materials and Devices») pp 1-50

[5] Malic B *et al* 2005 Alkaline-earth doping in $(K,Na)NbO_3$ based piezoceramics *Journal of the European Ceramic Society* **25** pp 2707–2711

[6] Nagaenko A V, *et al* 2020 Multi-element ferroactive materials based on KNN-PZT compositions with fundamentally different physical properties *Heliyon* **6** pp 3497

[7] Tomashpolsky Yu Ya, Matyuk V M, Sadovskaya N V 2016 Formation of a Surface Phase with the Matrix Composition in a $Ca_2Nb_2O_7$ Single Crystal *Inorganic Materials* **52** pp 791–795

[8] Zubarev J Y *et al* 2019 Intercalation of water molecules from the air into perovskite and layered structures formed in the system of $NaNbO_3$–$Ca_2Nb_2O_7$ *Heliyon* **5** pp e02786

[9] Titov S V *et al* 2000 Clustering of the structure preceded by concentration phase transitions *Letters in J.Tech. Phys* **26** pp 9–16

[10] Rao C N R, Gopalakrishnan J, 1990 New Directions in Solid State Chemistry. Structure, synthesis, properties, reactivity and materials design *New York. Cambridge university Press* pp 519

[11] Vezzoli G G 1982 Electrical properties of NbO_2 and Nb_2O_5 at elevated temperature in air and flowing argon *Phys.Rev. B* **26** pp 3954-3957

IOP Conf. Series: Materials Science and Engineering 1093 (2021) 012010 doi:10.1088/1757-899X/1093/1/012010

Numerical simulation of slow deformation perturbations in fault zones

A A Kazakbaeva[1] and I Yu Smolin[1,2]

[1]Tomsk State University, 36 Lenin ave., Tomsk, 634050, Russia
[2]Institute of Strength Physics and Materials Science of Siberian Branch of Russian Academy of Sciences, 2/4 Akademicheskii pr., Tomsk, 634055, Russia

E-mail: aigerim_@bk.ru

Abstract. Over the past half a century, the concept of slow deformation waves of the Earth has been developed and widely discussed in the Earth sciences. The velocity of slow waves is considered to be 5–6 orders of magnitude less than the velocity of sound and 7–8 orders of magnitude greater than tectonic flows. Analyzing and classifying various manifestations of slow deformation waves in the geomedium accumulated over forty years V. G. Bykov identifies two types of autowaves in his review—inter-fault and intra-fault. Here, the process of generation and propagation of slow deformation disturbances between two faults in an elastoplastic medium is studied numerically. The faults were defined as narrow elongated soft areas inclined to the axis of the load application. Just in these faults plastic deformation was permitted to generate. The simulations proved that the fronts of deformation waves head towards each other at approximately the same velocities, their shapes being close to planar but slightly curved.

1. Introduction

The study of the spatiotemporal migration of modern geodynamic processes is, on the one hand, one of the most interesting problems of geodynamics, on the other, one of the most debatable problems. Explanation of migration of earthquake foci in zones of large seismically active faults by the trigger effect of unidirectional spatial movement of slow inelastic deformations proposed by C. Richter [1] and K. Mogi [2] initiated the problem of slow deformation waves in the lithosphere.

Slow deformation waves are not directly recorded. Their existence is established indirectly by registration of variations in geophysical fields, in particular, by directional migrations of earthquakes. Involvement in the description of slow deformation perturbations, for example, heat conduction equations (diffusion) or the sin-Gordon equations, in the case of their representation as solitons, is based only on a priori certitude that such wave perturbations exist and are, for example, solitons. These equations are not connected to the evolution of the stress-strain state, which means that their solutions can only characterize a possible qualitative picture of slow deformation perturbations in geomedia. The autowave or autosoliton interpretation of these phenomena as self-organization processes in active dissipative media seems to be more adequate [3].

From the recent reviews presented in the papers [4, 5], it follows that there are two main types of deformation waves. Waves of the first type—inter-fault waves—are basic, have a long length and can originate in subduction zones, or be generated by the "lithosphere – asthenosphere" system, or have a different global nature. Waves of the second type—intra-fault—are generated first due to their transformation when passing through large inland fault zones. The velocities of spatial migration of inter-fault and intra-fault deformation waves differ and, according to different estimates, range from 10 to 100 km/year for the former and from 10 to 4 km/year or less for the latter [3].

Thus, an important task is to study the features of generation and interaction of slow deformation autowaves using the mathematical and numerical apparatus of deformable solid mechanics. A successful solution to this problem is possible on the basis of the fundamental theory of the evolution of the stress-strain state in a loaded solid medium, which covers the entire spectrum of key processes of deformation and fracture [6].

The aim of this work is to study the features of the formation of slow deformation perturbations in an elastoplastic medium using numerical simulation methods.

2. Mathematical model

To describe the processes of generation and propagation of slow deformation disturbances, in this work, we used a model that combines the approach of solid mechanics with cellular automata algorithm that provides a way for the plastic strain to advance into the elastic domain [7]. A model of an elastic-plastic media with the Drucker-Prager yield criterion was used. An algorithm of cellular automata with the von Neumann neighborhood was implemented for transmitting slow perturbation.

The complete mathematical model contains the fundamental equations of continuum mechanics (1)–(2), the decomposition of the stress tensor to the hydrostatic pressure and deviator stresses (3), constitutive evolutionary equations of an elastic–plastic medium (4)–(7) with unassociated plastic flow rule (8) and the Drucker-Prager plastic potential (9) and yield function (10).

$$\frac{d\rho}{dt} + \rho \operatorname{div}\bar{v} = 0 \tag{1}$$

$$\rho\frac{dv_i}{dt} = \frac{\partial \sigma_{ij}}{\partial x^j} + \rho F_i \tag{2}$$

$$\sigma_{ij} = -P\delta_{ij} + S_{ij};\ P = -\frac{1}{3}\sigma_{ii} \tag{3}$$

$$\dot{P} = -K(\dot{\varepsilon}^e_{ii}) \tag{4}$$

$$\dot{S}_{ij} = 2\mu\left(\dot{\varepsilon}^e_{ij} - \tfrac{1}{3}\dot{\varepsilon}^e_{kk}\delta_{ij}\right) - S_{ik}\dot{\omega}_{kj} + S_{kj}\dot{\omega}_{ik} \tag{5}$$

$$\dot{\varepsilon}^e_{ij} = \dot{\varepsilon}^t_{ij} - \dot{\varepsilon}^p_{ij} \tag{6}$$

$$\dot{\varepsilon}^t_{ij} = \frac{1}{2}\left(\frac{\partial v_i}{\partial x^j} + \frac{\partial v_j}{\partial x^i}\right);\ \dot{\omega}_{ij} = \frac{1}{2}\left(\frac{\partial v_i}{\partial x^j} - \frac{\partial v_j}{\partial x^i}\right) \tag{7}$$

$$\dot{\varepsilon}^p_{ij} = \dot{\lambda}\frac{\partial g(\sigma_{ij})}{\partial \varepsilon_{ij}},\ \text{if } f(\sigma_{ij}) \geq 0;\ \text{otherwise } \dot{\varepsilon}^p_{ij} = 0 \tag{8}$$

$$g(\sigma_{ij}) = \Lambda P + \sqrt{\frac{1}{2}S_{ij}S_{ij}} + const \tag{9}$$

$$f(\sigma_{ij}) = \alpha P + \sqrt{\frac{1}{2}S_{ij}S_{ij}} - Y \tag{10}$$

Here ρ is the density, v_i are the components of the velocity vector, σ_{ij} are the components of the stress tensor, F_i are the components of the vector of the bulk forces, P is the pressure, S_{ij} are the components of the deviatoric stress tensor, δ_{ij} is the Kronecker delta. K is the bulk modulus of elasticity, μ is the shear modulus, $\dot{\varepsilon}_{ij}^{e}$ and $\dot{\varepsilon}_{ij}^{p}$ are the elastic and plastic components of the total strain rate $\dot{\varepsilon}_{ij}^{t}$, respectively, the dot above the symbol means the material time derivative, $S_{ik}\dot{\omega}_{kj} + S_{kj}\dot{\omega}_{ik}$ is the correction for the rotation that occurs when using corotational Jaumann derivative of the stress tensor, $f(\sigma_{ij})$ is the yield function, $g(\sigma_{ij})$ is the plastic potential function, Λ is the dilatancy coefficient, α is the internal friction coefficient, $\dot{\lambda}$ is the plastic multiplier.

To perform numerical simulation, the Wilkins [8] finite difference scheme was used. Simulations were fulfilled using a computer code written by the authors. The computer code has been used previously for simulating various ductile and brittle materials, rocks, geomedia [9–13].

3. Results and discussion

The behavior of a region of the medium with two faults in the conditions of uniaxial compression along the vertical axis was studied. A structural map of the model media with faults was constructed to study the generation and propagation of slow strain waves between the faults. The model map has two narrow elongated areas specified as faults inclined to the axis of load application, their ends being placed near the boundary of the calculation region (figure 1a). Just the faults are the place where the plastic deformation can arise according to the idea of cellular automata.

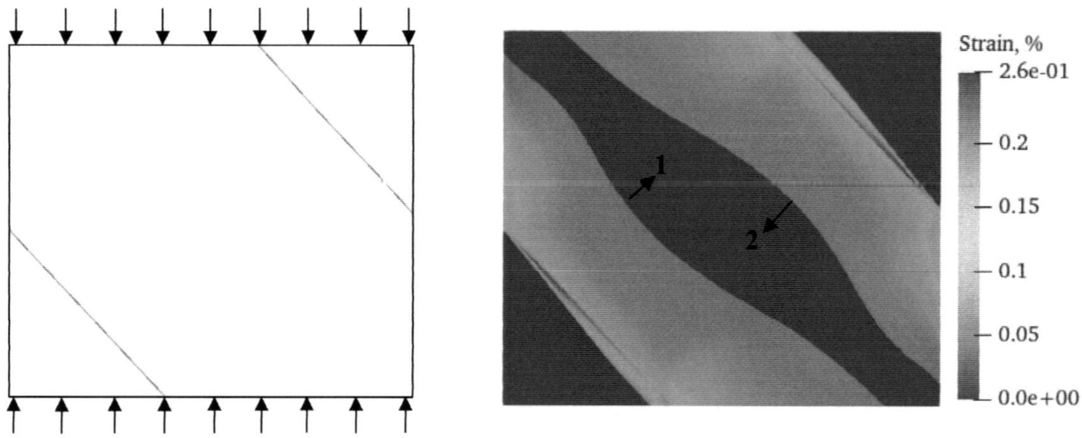

(a) (b)

Figure 1. Example of structural map of a model medium with faults and loading conditions (a), distribution of plastic strains (faults: number 1 and number 2) (b).

The simulations show that plastic strains generated in the faults and started to stem from the faults in opposite directions perpendicular to the faults (figure 1b). The fronts propagating toward the other fault have an almost planar slightly curved shape and approximately the same speed.

Figure 2 shows the distributions of increments of plastic strain that reflect the propagation of slow deformation perturbations in an elastic-plastic medium as a kind of solitary dissipative autowave. A detailed study showed that the speed of propagation of slow deformation perturbations exceeds the rate of loading of the boundaries of the calculated region by about 300 times. Taking into account the time compression [6], the average speed of propagation of slow deformation autowave number 1 (figure 1, b) was 27 km/year, and autowave number two (figure 1, a) equaled 23 km/year.

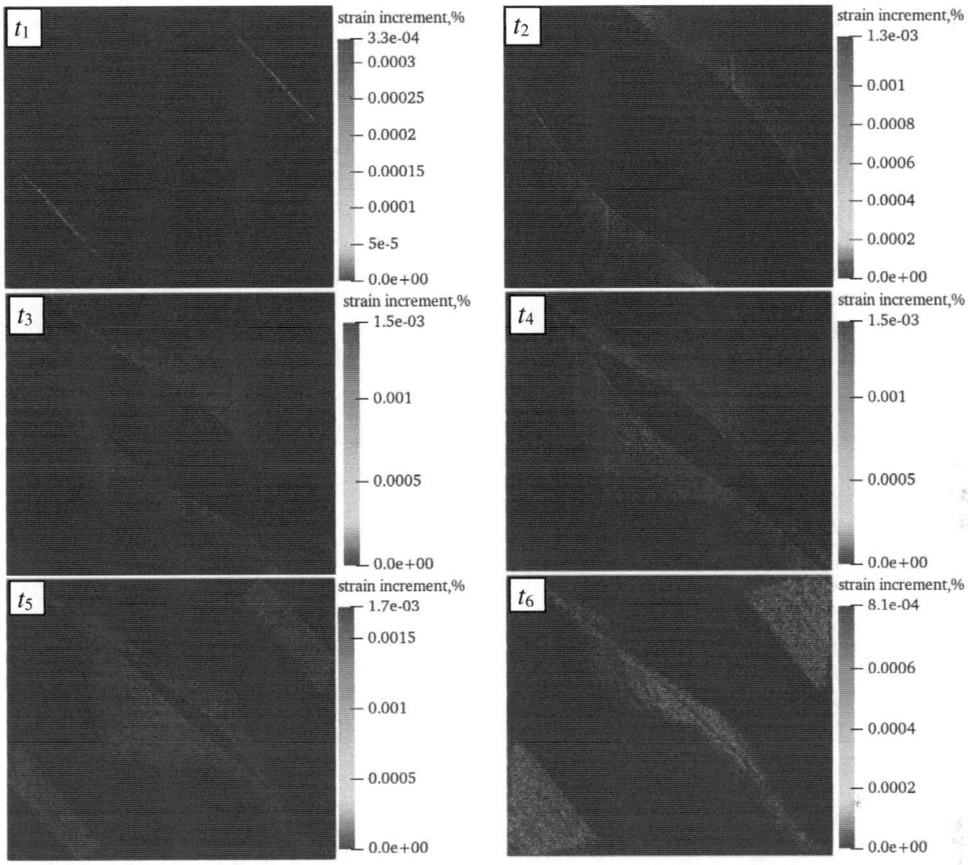

Figure 2. Distributions of strain increments of slow deformation waves.

4. Conclusions

As a result of the research, we can conclude that the approach based on the combined model of solid mechanics with the cellular automaton method allows one to describe slow deformation fronts of inelastic nature in nonlinear elastic-plastic media. The features of the propagation of deformation perturbations waves are investigated for a media with two elongated faults. The deformation perturbation fronts have an almost planar slightly curved shape and approach each other at approximately the same speed. The distributions of increments of plastic strain reflect the propagation of slow deformation perturbations in an elastic-plastic medium as a kind of solitary dissipative autowave. Variation of these distributions with time shows that the autowaves do not move steadily, their speed and width vary in time.

Acknowledgments

The study was carried out with the financial support of the Russian Science Foundation (project No. 19-17-00122).

References

[1] Richter E F 1958 *Elementary Seismology* (San Francisco: W. H. Freeman and Company) p 768
[2] Mogi K 1968 *Bull. Earthq. Res. Inst.* **46** (1) 53–74
[3] Kuzmin Yu O 2012 *Izv. Phys. Solid Earth* **48** (1) 1–16
[4] Bykov V G 2005 *Russ. Geol. Geophys.* **46** (11) 1158–70
[5] Sherman S I 2013 *Geodyn. Tectonophys.* **4** (2) 83–117

[6] Makarov P V, Khon Yu A and Peryshkin A Yu 2018 *Geodyn. Tectonophys.* **9** (3) 755–69
[7] Kazakbaeva A A, Smolin I Yu and Makarov P V 2020 *J. Phys.: Conf. Ser.* **1611** (1) 012067
[8] Wilkins M L 1999 *Computer Simulation of Dynamic Phenomena* (Berlin, Heidelberg: Springer-Verlag) p 246
[9] Mikushina V A 2020 *J. Phys.: Conf. Ser.* **1527** 012022
[10] Makarov P V, Smolin I Yu, Prokopinsky I P and Stefanov Yu P 1999 *Int. J. Fract.* **100** 121–31
[11] Bakeev R A and Stefanov Yu P 2019 *AIP Conf. Proc.* **2167** 020024
[12] Kulkov A S, Smolin I Y and Mikushina V A 2018 *AIP Conf. Proc.* **2051** 020162
[13] Garavand A, Stefanov Y P, Rebetsky Y L, Bakeev R A and Myasnikov A V 2020 *Int. J. Numer. Anal. Methods Geomech.* **44** 823–50

Deformation analysis of cylindrical lattice body of spacecraft in zone of high-pressure tank attachment

A A Khakhlenkova[1,2], **A E Burov**[1] **and A V Lopatin**[1,2]

[1]Federal Research Center for Information and Computational Technologies, Krasnoyarsk 660049, Russia
[2]Reshetnev Siberian State University of Science and Technology, Krasnoyarsk 660037, Russia

E-mail: sparklea@yandex.ru

Abstract. Finite element analysis of the strain-stress state in a load-bearing shell of the spacecraft body under longitudinal and lateral inertia loads was carried out. The shell is a composite anisogrid structure constituted by circumferential and helical ribs made of carbon fiber composite. A massive fuel storage tank is fixed inside the shell by means of a cable-stayed system. The shell structural behaviour in the zone of fuel tank installation was particularly addressed. A local reinforcement with additional circumferential ribs was shown to remarkably reduce the stress level and deflection in the tank attachment zone.

1. Introduction

Anisogrid lattice shells made of composite materials with superior specific strength and stiffness are widely used in rocket and space technology as they provide high weight efficiency and corresponding enhanced payload capabilities [1]. These structures are constituted by intersecting axial, circumferential and helical unidirectional composite ribs arranged in a repeating pattern that defines the membrane and bending stiffness. The lattice design concept provides a qualitative improvement in structural performance (particularly in buckling resistance) and design flexibility unachievable by continuous material counterparts. Although the principal manufacturing, design and analysis technologies for composite lattice structures are quite well developed [2-5], their mechanical behavior in each particular application still requires investigation.

Modern telecommunications satellites built on a modular design, such as those based on the Express platform, include a composite lattice cylinder as part of the service systems module. The shell is the primary load-bearing structure of the spacecraft body to which devices and mechanisms are attached. One of the main elements of the spacecraft propulsion system, a high pressure fuel storage tank, is fixed inside the shell structure by means of a cable-stayed system (figure 1). The tank is a metal-lined composite overwrapped pressure vessel made by filament winding [6-7]. The vessel designed to hold 300 l of xenon weighs about 440 kg at full load. During the launch phase of the mission, the spacecraft body experiences significant longitudinal and lateral overloads that are mandatory considered in design of the load-bearing shell structure. However, even if the overall (global) strength and stiffness requirements are met, the occurrence of local failure modes must be also evaluated especially in places where a stress gradient or/and massive equipment units are attached. Therefore, it is of considerable interest to analyze the structural behavior of the composite anisorgid shell near the installation of fuel tank. The results of such analysis performed using finite element simulation are presented in this paper. A local reinforcement solution to decrease stress level and deflection of the shell wall is also addressed.

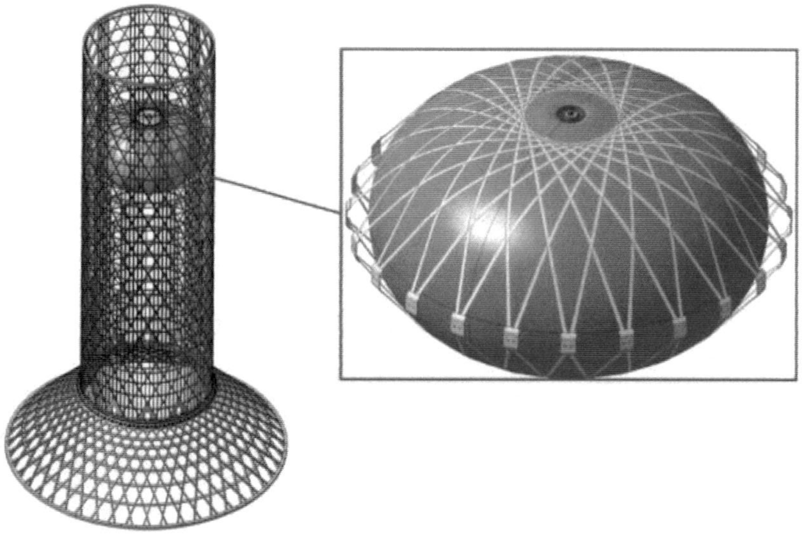

Figure 1. Composite anisorgid cylindrical shell of spacecraft body with attached fuel tank.

2. Finite element analysis

The lattice cylindrical shell under study with height $H = 3800$ mm and diameter $D = 1179$ mm consists of circumferential and helical ribs, and two end frames (figure 1). The helical ribs are oriented at $\varphi = \pm 14°$ angle with respect to the shell axis. The circumferential ribs pass through the middle of segments between the intersection points of helical ribs. The number of pairs of helical ribs $N = 60$. Helical and circumferential ribs have a rectangular cross-section of 15 x 6 mm and 15 x 2 mm, respectively. The end frames made of B95 aluminium alloy have a rectangular cross-section of 22 x 26 mm. Inside the shell at a height of 2 m from the bottom, the fuel high-pressure tank with a mass of 440 kg is installed. The tank is attached to the shell at 20 points evenly spaced around the circumference. The mechanical properties of the materials used in calculation are shown in Table 1.

Table 1. Properties of materials used in calculation.

Anisogrid lattice	Modulus of elasticity along the rib E_1 (GPa)	180
	Modulus of elasticity along the rib height E_2 (GPa)	6.7
	Shear modulus G_{12} (GPa)	3.5
	Shear modulus G_{23} (GPa)	2.3
	Poisson's ratio μ_1	0.19
	Density ρ (kg/m^3)	1550
	Compressive strength along the rib S_{1c} (MPa)	480
	Tensile strength along the rib S_{1t} (MPa)	800
End frame	Modulus of elasticity E (GPa)	71
	Poisson's ratio μ	0.3
	Density ρ (kg/m^3)	2750

Numerical analysis on deformation behavior of the anisogrid lattice cylindrical shell was carried out using a three-dimensional finite-element model constructed using MSC Nastran package [8]. The shell structure is modelled with two-node BEAM type elements. The whole shell model is formed from FE model of repeating unit cell using operations of copying, rotating and transferring. The unit cell model (figure 2) consists of two fragments of helical ribs and a fragment of circumferential rib. The width b_c and height h_c of the unit cell are determined from the shell design parameters:

$$b_c = \frac{\pi D}{N} \qquad h_c = \frac{b_c}{2\tan\varphi} \tag{1}$$

The finite element model is built up using an integrated macro code. Initially, a finite element model of the unit cell is created (figure 2). At this stage, the geometric parameters and elastic properties of ribs as well as the size of elements are specified. Then, using the copying procedure, a finite element model of a ring of interconnected unit cells is built. The ring elements are mirrored followed by successive copying of the obtained elements along the shell axis until the given shell height is achieved. The circumferential ribs at the shell ends are assigned the properties of aluminum frames. The bottom end of the shell is fixed from vertical and horizontal displacement. The spacecraft structure is subjected to longitudinal and lateral overloads of 11g and 3.5g, respectively.

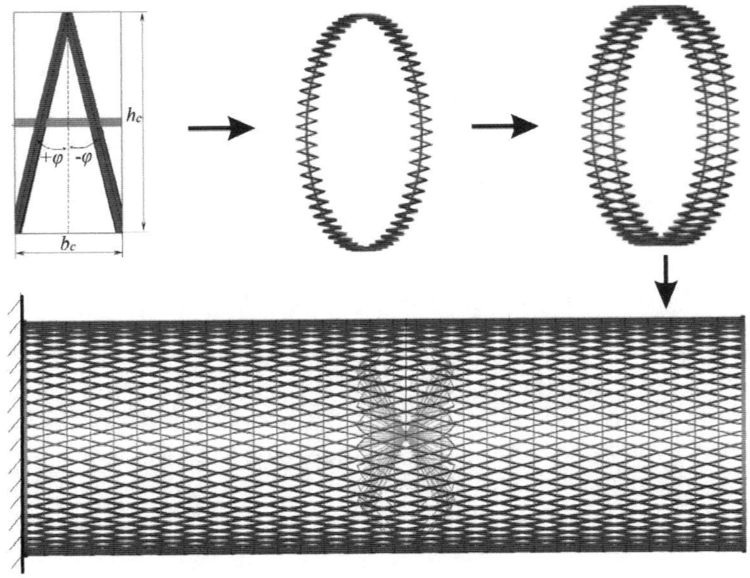

Figure 2. Building the finite element model of the shell.

The fuel tank is represented by a rigid element (RIGID type) with a concentrated mass element (MASS type) at the central node. The tank is connected to the shell wall using the beam elements that simulate a system of twenty uniformly tensioned cables. To evaluate the influence imposed by the cable-stayed system on deformation of the shell, a pretension P ranged from 500 to 2500 N with step of 250 N is set in these elements by means of the Bolt Preload tool. The cross-section and mechanical properties of the cable elements are selected so that the first natural vibration frequency of the tank structure matches the real value. It should be mentioned that the higher the pretention, the stiffer is the cable-stayed system. The rigid cable-stayed system is desirable as this improves the mechanical performance of the shell against the longitudinal acceleration.

3. Results and discussion

As expected, the maximum stress and deformation in the shell wall under applied loads occur in the zone of the installation of fuel tank and their magnitudes are governed by pretention in the cable-stayed system. There is a practically linear dependence of the minimum compressive stress (figures 3) and maximum deflection, which is inward of the shell (figure 4) on the pretention force. Here, the attention is paid for the compressive stresses as their ratio to the ultimate value is much higher than that for the tensile stresses. The compressive stresses reach the compressive strength of the rib material when the cable pretention is equal or exceed 1750 *N*.

More crucial in terms of structural integrity is local bending of the shell wall in the installation zone as ribs buckling can occur at a stress level which is much less than the ultimate compressive strength [9-10]. The deformation pattern shown in figure 5 evidences a significant local deflection of the shell wall caused by the combined action of lateral load and cable-stayed system. This can result in local buckling failure and overall reduction in the load carrying capacity of the shell.

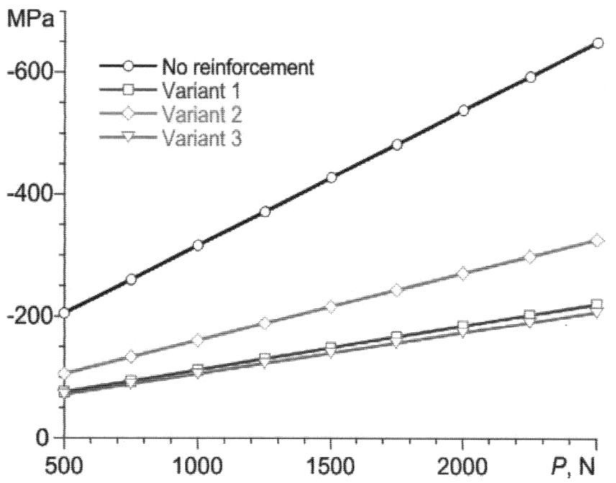

Figure 3. Minimum compressive stress depending on the cable pretention.

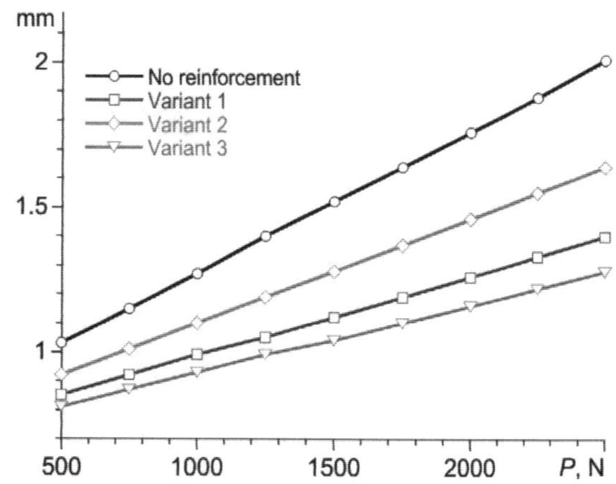

Figure 4. Maximum lateral deflection of the shell wall depending on the cable pretention.

There are several ways to improve the local buckling resistance of lattice structures including the use of denser lattice configuration or local reinforcement. However, the former will inevitably leads to a significant increase in mass and corresponding reduction of the weight efficiency of the shell structure. To decrease the stress level and local bending in the shell, consider the reinforcement of the tank installation zone with additional circumferential ribs. Three variants of the ribs placements shown in figure 5 were examined. Note that the width of additional ribs which highlighted in red on figure 5 is not in scale. All the proposed options of local reinforcement can be easily implemented in the manufacturing process of the shell.

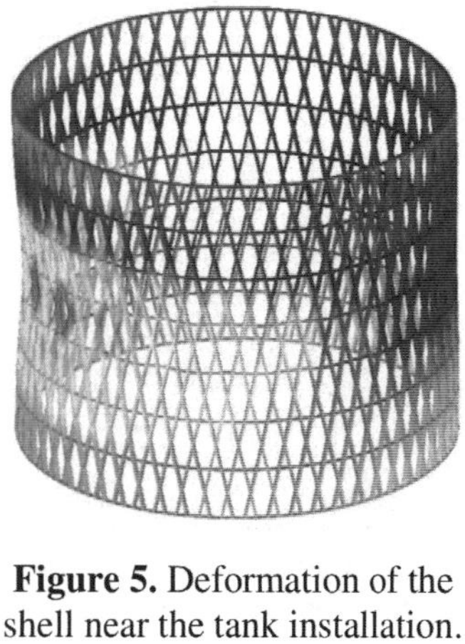

Figure 5. Deformation of the shell near the tank installation.

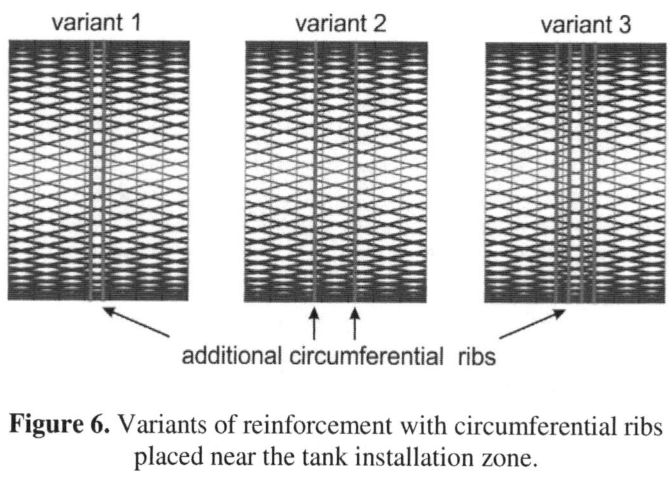

Figure 6. Variants of reinforcement with circumferential ribs placed near the tank installation zone.

As follows from figures 3 and 4, the use of all variants of reinforcement resulted in reduction in both the compressive stresses and deflection in the tank installation zone. Regardless of the reinforcement scheme, the maximum stress does not exceed the material compressive strength for entire range of pretension values. Interestingly, the reinforcement according to variant 1 demonstrates practically the same dependence of the minimum compressive stress on the cable pretension as that according to variant 3 (figure 3), although in the former case only one pair of additional ribs are used.

The reinforcement variant 1 is preferable to variant 2 as it provides a greater reduction in the local bending of the shell wall. The most effective in terms of reducing the lateral deflections turned out to be the variant 3 of reinforcement. It is not surprising as this reinforcement scheme is in fact a combination of variant 1 and variant 2. The reinforcement solution allows to reduce the deflection of the shell wall by 27 % at P = 500 N and by 57 % at P = 2500 N in comparison with the initial unreinforced structure.

4. Conclusion

The composite anisorgid cylindrical shell of the spacecraft body with the fuel tank attached by means of cable-stayed system under longitudinal and lateral overloads was numerically analysed. The results revealed that there are the elevated stress level and large lateral deflection of the shell wall near the tank installation zone. The local reinforcement with additional circumferential ribs remarkably reduced the stress and deflection. This improved the strength and local buckling resistance of the shell structure. Different variants of such kind of reinforcement can be implemented in the manufacturing process of the shell. The presented approach can be applied in design of composite anisogrid structures used in the aviation and space applications.

Acknowledgments
This work was supported by the Ministry of Science and Higher Education of the Russian Federation, grant RFMEFI60419X0233.

References

[1] Vasiliev V V and Razin A F 2006 Anisogrid composite lattice structures for spacecraft and aircraft applications *Compos. Struct.* **76** 182-189
[2] Vasiliev V V, Barynin V A and Razin A F 2012 Anisogrid composite lattice structures – Development and aerospace applications *Compos. Struct.* **94** 1117-1127
[3] Zheng Q, Jiang D Z, Huang C F, Shang X L and Ju S 2015 Analysis of failure loads and optimal design of composite lattice cylinder under axial compression *Compos. Struct.* **131** 885-894
[4] Beerhorst M and Hühne C 2016 Optimization of axially compressed cylindrical grid structures using analytical and numerical models *Compos. Struct.* **157** 155–162
[5] Shroff S, Acar E and Kassapoglou C 2017 Design, analysis, fabrication, and testing of composite grid-stiffened panels for aircraft structures *Thin-Walled Struct.* **119** 235-246
[6] Vasiliev V V 2009 *Composite Pressure Vessels - Analysis, Design, and Manufacturing* (Blacksburg, Virginia: Bull Ridge Publishing)
[7] Burov A E, Lepihin A M, Makhutov N A and Moskvichev V V 2017 Numerical analysis of stress-strain state and strength of metal lined composite overwrapped pressure vessel *Strength Mater.* **19** 666-675
[8] MSC Software Corporation 2014 *MSC Nastran Quick Reference Guide*
[9] Totaro G 2013 Local buckling modeling of isogrid and anisogrid lattice cylindrical shells with hexagonal cells *Compos. Struct.* **95** 403-410
[10] Totaro G, Nicola F De and Caramuta P 2013 Local buckling modelling of anisogrid lattice structures with hexagonal cells: An experimental verification *Compos. Struct.* **106** 734-741

Influence of grain size on the nucleation and development of plasticity in nanocrystalline FeNi films

A V Korchuganov, D S Kryzhevich, Y A Chumakov and A S Grigoriev

Institute of Strength Physics and Materials Science SB RAS, 2/4 Akademicheskii ave., Tomsk, 634055, Russian Federation

E-mail: avkor@ispms.ru

Abstract. A molecular dynamics study of the features of the nucleation and evolution of plastic deformation in nanosized nanocrystalline FeNi films under uniform uniaxial tension is carried out. The dependences of the strength properties of the films on the grain size are obtained. It is found that the dependences of stresses at which defects begin to nucleate, maximum strength is achieved and plastic flow is realized, on the grain size have a pronounced maximum. The optimal grain structure for which the nanocrystalline film has the maximum strength at uniaxial stretching is determined. The calculated dependences allow determining the critical grain size at which the Hall-Petch relationship is reversed. It is shown that the nucleation and initial development of plasticity in nanosized films with grain size above the critical one has a dislocation nature. The twinning contribution to the film plasticity increases at strains corresponding to plastic flow.

1. Introduction

The grain size is an important material characteristic that determines the activation or suppression of various mechanisms of plastic deformation. As a rule, as the grain size decreases to hundreds or tens of nanometers, the strength of material increases in comparison with coarse-grained counterparts. It is known that the onset of plasticity in metallic coarse-grained materials has a dislocation nature. In nanocrystalline metals with a grain size less than 20-30 nm, the dislocation mobility is significantly suppressed, since the Peierls barrier, which prevents dislocation slip, is sufficiently high [1]. Therefore, in nanocrystalline materials, along with dislocation processes, various grain boundary deformation mechanisms [1-3] or twinning [4-6] are activated. One of the main grain boundary processes is grain boundary sliding, which, in particular, prevails during superplastic deformation of nanocrystalline metals [7]. The important role of grain boundary sliding in the plasticity of nanomaterials at high stresses and strain rates is also confirmed by computer simulations [8,9]. Grain boundary sliding is realized in the form of plastic shears localized in the region of grain boundaries. They lead to the formation of structural defects at triple junctions, which become sources of internal stresses [10]. In turn, these defects can lead to the nucleation of nanocracks and subsequent brittle fracture of the material. Simultaneously with grain boundary sliding, accommodation processes can be initiated in the material, which are caused by the transformation of defects formed in triple junctions due to stress relaxation localized there. This leads to an increase in the ductility of the material. Revealing the nature of accommodation of grain boundary sliding in nanocrystalline materials is of great scientific and practical interest. The mechanisms of accommodation of grain boundary sliding in such materials can be the emission of lattice dislocations from triple junctions [7], twinning [11,12], migration and splitting of grain boundaries [13,14], rotational deformation [15]. This behavior during deformation is typical for bulk nanocrystalline materials. However, in nanoscale nanocrystalline films, the mechanisms of nucleation and development of plasticity can differ, which is caused by a large fraction of the material located near free surfaces.

The leading role of particular accommodation process is associated with the grain size in the nanocrystalline material. The strength characteristics of nanocrystalline metallic materials increase with a decrease in the grain size only up to a certain density of grain boundaries, which is well described by the Hall-Petch relationship [16]. Starting from a certain critical grain size, further refinement of the grain structure leads to a softening of the material. For such materials, strength and grain size are related by the inverse Hall-Petch relationship. The physical nature of the inverse Hall-Petch relationship is still a subject of discussions [16-18].

The transience and small spatial scales of the processes that determine the behavior of materials at the atomic level impose difficulty on experimental investigations. At the same time, computer simulation allows one to study in detail the features of the nucleation and development of structural rearrangements under mechanical loading including high-energy action on the material [19-24].

In this work, the features of the nucleation and development of plasticity in nanosized nanocrystalline FeNi films under uniaxial uniform tension were investigated. We studied the effect of grain size on the mechanical characteristics of the samples to determine the optimal grain structure, at which the simulated films have the maximum strength under uniaxial tension.

2. Methods

The simulation was carried out on the base of the molecular dynamics method using the LAMMPS software package [25]. The interatomic interaction in the Fe-Ni system was described by the many-body potential developed in the framework of the embedded atom method [26]. The chemical composition of the samples corresponded to the equiatomic composition: Fe50%-Ni50%. The loaded samples consisted of 8 grains with approximately the same size. The centers of the grains were located at the sites of the conditional hcp lattice. The Voronoi tessellation was used to construct the grains. The grain sizes varied from 5 to 20 nm. For each grain size, 10 samples were loaded with different misorientation angles between the grains. The sample sizes, depending on the grain size, varied in the range from 10×8×8 nm to 40×32×32 nm. The corresponding number of atoms in the samples ranged from 63 000 to 4 080 000. The simulated sample with a grain size of 20 nm is shown in figure 1. Periodic boundary conditions were used along the *X* and *Y* directions, and free surfaces were specified in the *Z* direction. The simulated samples were nanosized films, with two layers of grains parallel to the free surfaces. Uniaxial tension of the film was specified in the form of uniform stretching along the *X* direction at a constant rate of 5 m/s by means of a corresponding increase of the *X* coordinate for all atoms.

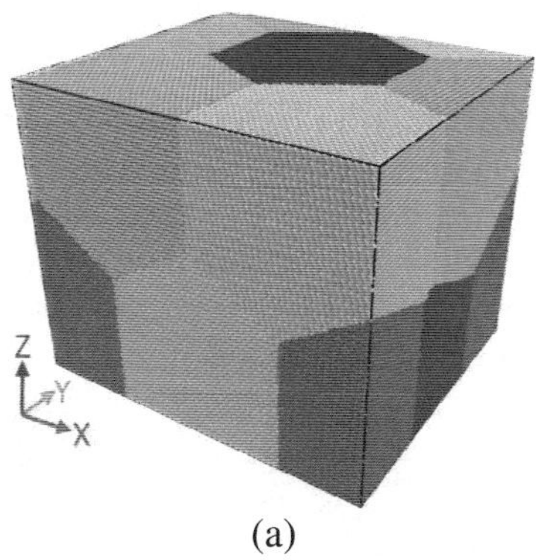

(a)

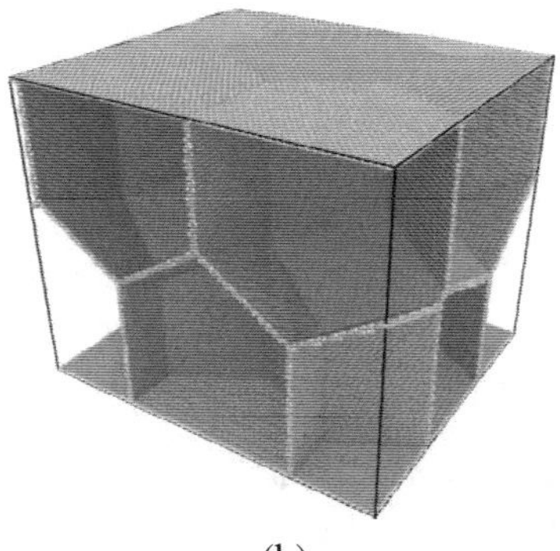
(b)

Figure 1. The initial structure of the nanocrystalline FeNi film with 20 nm grain size. Each grain is highlighted in a different color in figure (a). Only atoms on free surfaces and at grain boundaries are shown in figure (b).

During relaxation and loading of the samples, the thermostat rescaled a temperature to 300 K, and barostat maintained zero stress in the direction of the *Y* axis. The identification of structural defects in loaded samples was carried out using the Common Neighbor Analysis (CNA) algorithm [27]. The structure of the sample was visualized in the OVITO software package [28].

3. Results and discussion

Let us consider the most typical processes of nucleation and development of plasticity by the example of a nanocrystalline film with 20 nm grain size. Analysis of the simulation results showed that the features of the dependences of stress and the fraction of atoms with HCP symmetry of the nearest environment on the strain, shown in figure 2, correlate quite well with the features of the change in the defect structure of the loaded film (figure 3). Note that Shockley partial dislocations moving along slip planes leave intrinsic stacking faults (ISFs) behind. They are identified by the CNA algorithm as two layers of atoms with HCP symmetry of the nearest environment (hereinafter – HCP atoms) [29]. According to this algorithm, twin boundaries represent one layer of HCP atoms. Therefore, the fraction of HCP atoms in figure 2b is a rather informative characteristic that reflects the total area of nucleated defects.

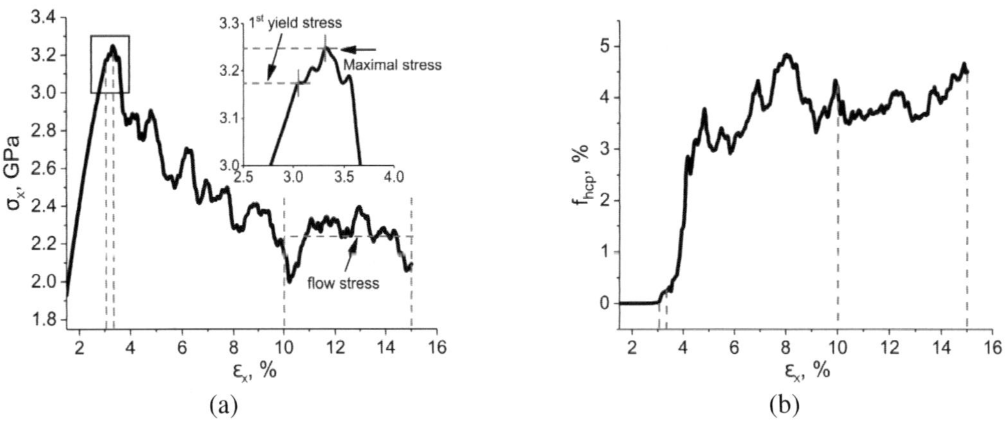

Figure 2. Dependences of the stress along the *X* axis (a) and the fraction of HCP atoms (b) on the strain of a nanocrystalline FeNi film with 20 nm grain size.

The first maximum on the σ(ε) curve in figure 2a will be called first yield stress. When the deformation corresponding to this maximum is reached, the first partial dislocation is generated and propagated in one of the grains of the film, leaving ISF behind which is highlighted in red in figure 3a. This leads to the gradual increase in the fraction of HCP atoms. The nucleated dislocation propagates through the grain body and stops at a neighboring interface, which can be either a grain boundary or a free surface. At further increase in strain to 3.30%, an absolute maximum (maximal stress) appears on the σ(ε) curve. Starting from this strain, the nucleation and propagation of numerous structural defects takes place in the simulated film. This corresponds to an avalanche-like increase in the fraction of HCP atoms up to several percent (figure 2b). Such significant changes in the defect structure of the film lead to a remarkable decrease in stress (figure 2a). The defect structure corresponding to 4.25% strain is shown in figure 3b. After reaching this strain, partial dislocations had glided through almost all grains, leaving ISFs behind. Some ISFs, as a result of gliding of partial dislocations in a plane adjacent to them, are transformed into twins with a thickness of two atomic planes, which are highlighted in yellow in figure 3. When the simulated film is stretched in the interval of 4.50-6.25%, some ISFs "disappear" (figure 3c). This is due to trailing partial dislocations 1/6<112>{111} glide in the ISF plane, which lead to a shear corresponding to the full dislocation 1/2<110>{111}, and restore an ideal FCC lattice. Approximately a half of all ISFs are transformed into thin twins simultaneously with the above process. When the sample is stretched in the interval of 6.25-8.00%, new ISFs are nucleated (figure 3d).

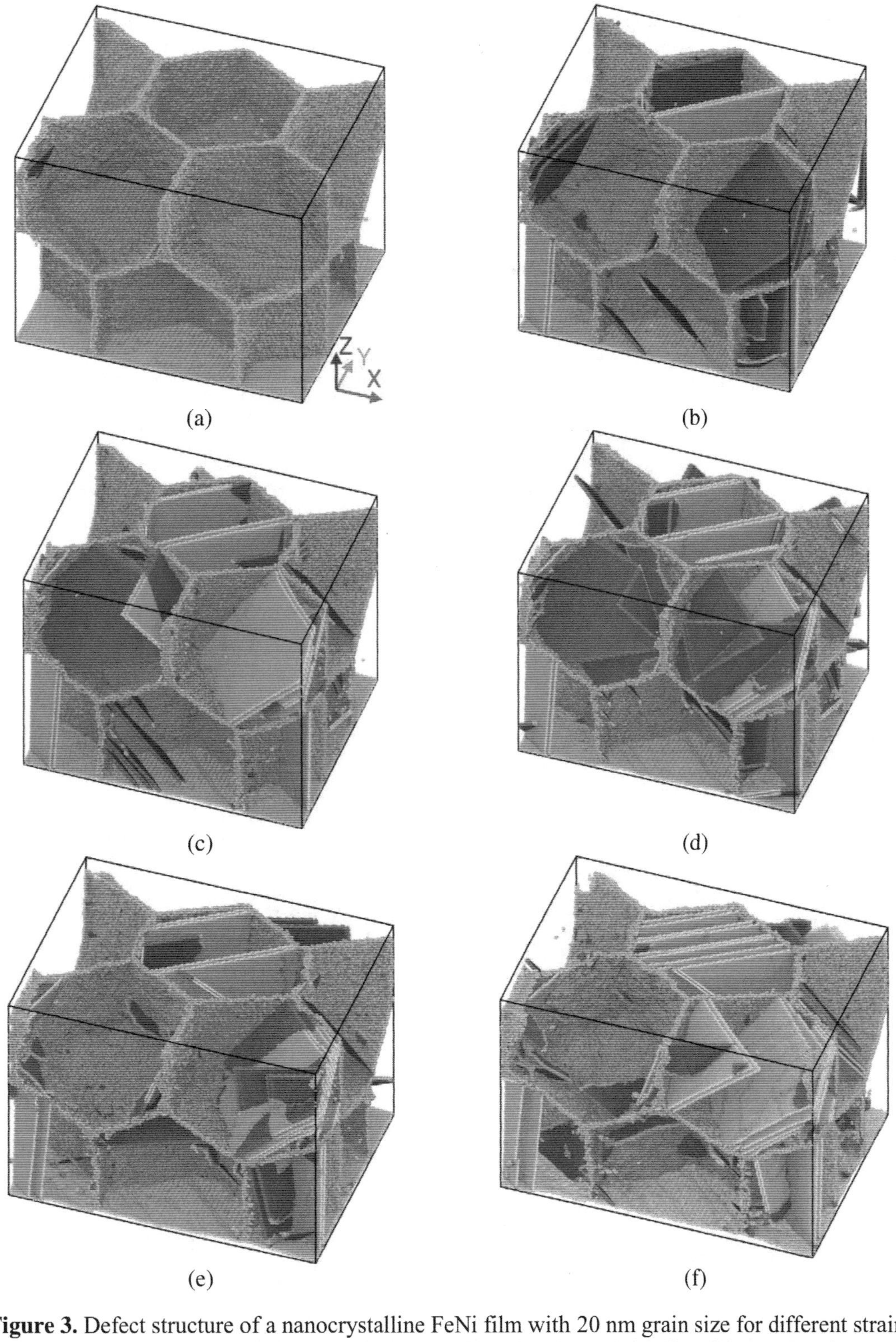

Figure 3. Defect structure of a nanocrystalline FeNi film with 20 nm grain size for different strain values: 3.10% (a); 4.25% (b); 6.25% (c); 8.00% (d); 10.50% (e); 15.00% (f). FCC atoms and atoms on the upper free surface are not shown. Atoms that belong to grain boundaries or free surfaces, intrinsic stacking faults and twin boundaries are colored in gray, red and yellow, respectively.

Then, up to 10.50% strain, the trailing dislocations glide along the ISFs and decrease the ISF area (figure 3e). As a result, some grains become almost defect-free. At the same time the area of twin boundaries remains virtually unchanged. Despite the fact that the fraction of HCP atoms for 10.50% strain is virtually the same as for 6.25%, in the interval between these strains a large number of split full dislocations glided in the grains, providing a significant decrease in stress. The stage of plastic flow begins at further deformation, at which the stresses and the fraction of HCP atoms change only slightly. This is due to the fact that twins begin to increase their thickness through the sliding of partial dislocations in adjacent planes at this stage of plastic deformation (figure 3f). Since the twin boundaries simply migrate to the adjacent plane in this case, the number of HCP atoms does not change. Almost all ISFs are transformed into twins. At the same time, grains without twins undergo multiple shears due to successive glide of Shockley leading and trailing dislocations.

As noted earlier, the grain size is the most important parameter of the internal structure that determines the physical and mechanical properties of the material. Therefore, it was of interest to reveal the effect of the grain size on the stresses at which the first yield stress occurs, the maximal stress is reached, and the plastic flow of the material is realized. In this case, the flow stress was calculated by averaging the stress in the interval of strains from 10 to 15%. The calculation results showed that the dependences of the first yield stress, maximal stress and flow stress on the grain size are significantly nonlinear and have a pronounced maximum. Note that the curves of dependences of first yield stress and maximal stress on average grain size in figure 4a correlate well with each other. FeNi nanocrystalline films with an average grain size of 12 nm are characterized by the highest strength with respect to the nucleation of structural defects. The critical grain size for the flow stress of nanocrystalline films is in the interval from 8 to 10 nm (figure 4b). This means that the flow stresses in nanocrystalline films with average grain sizes less than 8 and more than 10 nm are described by the inverse and classic Hall-Petch relationships, respectively.

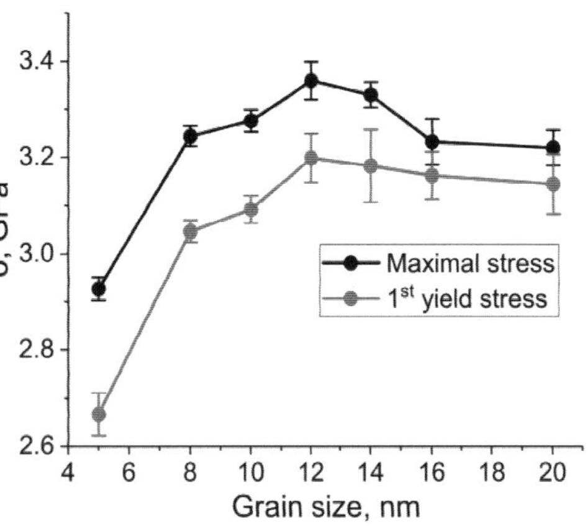

(a)

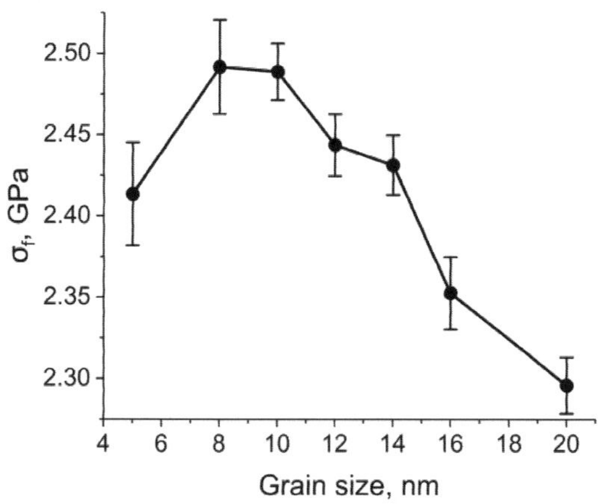

(b)

Figure 4. Dependences of the average values of the maximum stress (black curve), first yield stress (red curve) (a) and flow stress (b), on the grain size in nanocrystalline FeNi films.

4. Conclusions

Based on the simulations performed, it can be concluded that the nucleation and development of plasticity in nanoscale films with a nanocrystalline structure under uniaxial tension has a dislocation nature. As the strain increases, twinning processes begin to play more important role in the deformation behavior of the films. It is shown that a decrease in the average grain size down to 8-10 nm increases the tensile strength of the films. This critical size defines the boundary at which the Hall-Petch relationship is reversed. At further decrease in the grain size, the strength of the films decreases. It is possible that this softening of the films is associated with the activation of intergranular processes, such as grain boundary sliding, grain boundary migration, grain rotations, etc.

Acknowledgment
The work was carried out with the financial support of Russian Science Foundation (project No. 20-79-10406).

References
[1] Ovid'ko I A 2007 *J. Mater. Sci.* **42** 1694–708
[2] Koch C C 2007 *J. Mater. Sci.* **42** 1403–14
[3] Wolf D, Yamakov V, Phillpot S R, Mukherjee A and Gleiter H 2005 *Acta Mater.* **53** 1–40
[4] Zhu Y T, Wu X L, Liao X Z, Narayan J, Mathaudhu S N and Kecsḱs L J 2009 *Appl. Phys. Lett.* **95** 031909
[5] Wu X L and Zhu Y T 2008 *Phys. Rev. Lett.* **101** 025503
[6] Zhu Y T, Liao X Z and Valiev R Z 2005 *Appl. Phys. Lett.* **86** 1–3
[7] Mukherjee A K 2002 *Mater. Sci. Eng.* A **322** 1–22
[8] Monk J, Hyde B and Farkas D 2006 *J. Mater. Sci.* **41** 7741–6
[9] Husain A, La P, Hongzheng Y and Jie S 2020 *Materials* **13** 3223
[10] Morozov N F, Ovid'ko I A, Petrov Y V and Sheinerman A G 2009 *Rev. Adv. Mater. Sci.* **19** 63–72
[11] Zhu Y T, Liao X Z, Wu X L and Narayan J 2013 *J. Mater. Sci.* **48** 4467–75
[12] Ahmed S I, Mkhoyan K A and Youssef K M 2020 *Mater. Sci. Eng.* A **777** 139069
[13] Wang L, Xin T, Kong D, Shu X, Chen Y, Zhou H, Teng J, Zhang Z, Zou J and Han X 2017 *Scr. Mater.* **134** 95–9
[14] Zolnikov K P, Kryzhevich D S and Korchuganov A V 2019 *Russ. Phys. J.* **62** 46–51
[15] Bobylev S V, Mukherjee A K and Ovid'ko I A 2009 *Rev. Adv. Mater. Sci.* **19** 103–13
[16] Cordero Z C, Knight B E and Schuh C A 2016 *Int. Mater. Rev.* **61** 495–512
[17] Naik S N and Walley S M 2019 *J. Mater. Sci.* **55** 2661–81
[18] Carlton C and Ferreira P J 2006 *Mater. Res. Soc. Symp. Proc.* **976** 19–24
[19] Zolnikov K P, Korchuganov A V, Kryzhevich D S, Chernov V M and Psakhie S G 2019 *Phys. Mesomech.* **22** 355–64
[20] Zolnikov K P, Kryzhevich D S and Korchuganov A V 2020 *Russ. Phys. J.* **63** 947–53
[21] Zolnikov K P, Korchuganov A V, Kryzhevich D S and Psakhie S G 2018 *Phys. Mesomech.* **21** 492–7
[22] Lychagin D V, Dmitriev A I, Nikonov A Y and Alfyorova E A 2020 *Crystals* **10** 666
[23] Shugurov A R, Panin A V, Dmitriev A I and Nikonov A Y 2020 *Metals* **10** 1332
[24] Dmitriev A I, Nikonov A Y, Shugurov A R and Panin A V 2021 *Mater. Sci. Eng.* A **800** 140327
[25] Plimpton S 1995 *J. Comput. Phys.* **117** 1–19
[26] Zhou X W, Foster M E and Sills R B 2018 *J. Comput. Chem.* **39** 2420–31
[27] Honeycutt J D and Andersen H C 1987 *J. Phys. Chem.* **91** 4950–63
[28] Stukowski A 2010 *Model. Simul. Mater. Sci. Eng.* **18** 015012
[29] Psakhie S G, Zolnikov K P, Kryzhevich D S and Korchuganov A V 2019 *Sci. Rep.* **9** 3867

IOP Conf. Series: Materials Science and Engineering 1093 (2021) 012013 doi:10.1088/1757-899X/1093/1/012013

Peculiarities of the development of plastic deformation in a textured FeNi alloy with a gradient grain structure

A V Korchuganov, D S Kryzhevich, Y A Chumakov and A S Grigoriev

Institute of Strength Physics and Materials Science SB RAS, 2/4 Akademicheskii ave., Tomsk, 634055, Russian Federation

E-mail: avkor@ispms.ru

Abstract. The mechanisms of nucleation and development of plastic deformation in the FeNi alloy with a gradient grain structure with a pronounced axial texture were revealed. The study was carried out on the basis of a molecular dynamics method. The simulated sample contained four layers, each of which consisted of grains of approximately the same size. The sample was subjected to uniform uniaxial tension at a constant rate in the direction of the grain misorientation axis. It was found that initially plastic deformation develops only in the layer with the largest grains and represents an abrupt growth of stacking faults. This process ends with the hardening of the deformed sample. At higher deformations, smaller grains contribute to the development of plasticity. At the same time, an almost ideal crystal structure can be restored in large grains, and twinning occurs in medium-sized ones.

1. Introduction

The development of new heterogeneous and hierarchical structures that combine high strength and plasticity is one of the most promising areas in materials science [1–3]. Among the wide variety of heterogeneous structures, materials with a gradient grain structure attract close attention of researchers all over the world for the uniqueness of their physical and mechanical properties [4-7]. In such materials, the grain size changes monotonically in one of the directions. As a rule, the grain size decreases towards the free surface of the samples. Such an ordering of grains by size leads to the formation of regions with significantly different mechanical properties in the material. The combination of these regions significantly improves the performance characteristics of the material.

The reasons for the simultaneous achievement of high strength and ductility in gradient materials are associated with the following processes. Initially, plastic deformation arises in coarse grains, and as the mechanical load increases, it gradually spreads to regions with smaller grains. As a result, there is a redistribution of stresses between grains of different sizes with different strengths in the sample, which leads to the suppression of deformation localization [8]. At further loading, delocalization of deformation gradually spreads to grains with smaller sizes until it reaches the smallest ones [9]. In this case, large deformed grains are strengthened due to the accumulation of dislocations, providing a general strain hardening of the sample [10]. At this time, small grains are softened by migration of grain boundaries and growth [11]. Thus, various mechanisms of plastic deformation are simultaneously activated in grains of different sizes in the gradient grain structure. This allows the surface layer of the gradient material to be deformed by a significantly greater value than a homogeneous nanocrystalline counterpart maintaining high strength characteristics of the whole sample.

Despite the general understanding of the plasticity processes, the features of various atomic mechanisms that determine the response of gradient materials under mechanical loading are largely unclear and require detailed study. The most detailed and deeply the above questions can be solved on the basis of computer simulation of the behaviour of materials at the microscopic level [12-14]. In this work we studied the features of the nucleation and development of plasticity in a gradient nanocrystalline FeNi sample with a pronounced axial grain texture under uniaxial tension on the basis of computer simulation.

2. Methods

The studies were conducted on the basis of the molecular dynamics method using the LAMMPS software package [15]. The interatomic interaction in the FeNi alloy was described by the many-body potential [16]. The initial temperature of the sample was set equal to 300 K. Nanocrystalline sample in the shape of a parallelepiped with a gradient grain structure was created. In the direction of the Z axis (figure 1), four layers were formed, in each of which the grains had approximately the same size: 15, 30, 15 and 10 nm. The sample contained 4.7 million atoms, its size was 30x25x70 nm in the *X*, *Y*, and *Z* directions respectively. The grain structure of the sample was built on the basis of space partitioning by the Voronoi-Laguerre tessellation method. The orientation of the grains was set by rotation around the [123] crystallographic direction, which coincides with the *Z* axis of the sample, by a random angle, while the angle between adjacent grains could not be less than 30 degrees. The sample was uniformly uniaxially stretched in the *Z* direction at a constant speed of 5 m/s. Figure 1 shows the structure and the orientation of the sample. The analysis of the simulation results was carried out on the basis of a complex of modern programs and algorithms for the identification and visualization of structural defects [17,18].

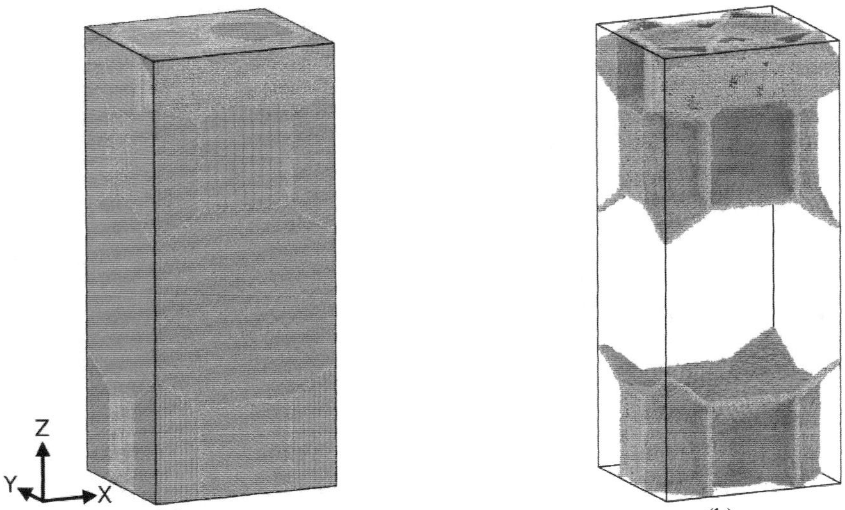

Figure 1. FeNi sample with a gradient grain structure before loading (a). Grain boundaries in the simulated sample (b).
Atoms in the bulk of grains and in the interface regions are marked in green and grey, respectively.

3. Simulation results

As shown in figure 1, all grains in the simulated crystallite before loading had an ideal crystal lattice, which was preserved up to a certain threshold strain value. This threshold value of strain at which structural defects begin to nucleate in the simulated samples can be determined quite accurately from the first kink in the stress-strain curve. As shown in figure 2a, the first kink (local maximum) on the stress-strain curve occurs when the sample is stretched by ~2.2%. At this deformation, an avalanche-like growth of intrinsic stacking faults (SF) begins in the largest grain located in the central part of the sample, which corresponds to an increase in the number of atoms with HCP symmetry of the nearest environment in figure 2b. At strains in the interval from 2.2 to 2.5%, the stress in the sample abruptly decreases by about 30–35%, and the fraction of atoms whose nearest environment has HCP symmetry increases to about 5%. These atoms initially appear at the grain boundaries of the largest grain as a result of the nucleation of Shockley partial dislocations. These dislocations glide in the coarse grain up to the opposite grain boundary and stop there. SFs are formed in the slip planes of partial dislocations. In figure 3a, it is clearly seen that multiple SFs are formed in the coarse grain, which provided structure accommodation of the whole sample for applied loading. At further stretching to about 3.9%, the stress in the hardened sample increases. In the strain interval from 2.6 to 3.9%, the fraction of atoms with HCP nearest environment does not change, and the corresponding curve in figure 2b reaches a plateau. This means that new SFs are not formed in the sample. When the sample is stretched in the range from 3.9 to 7.0%, plastic deformation develops in medium-sized grains. It is also realized by the generation and propagation of SFs in the grains of medium size (figure 3b) and annihilation of SFs by trailing dislocations in the largest grain (figure 3c).

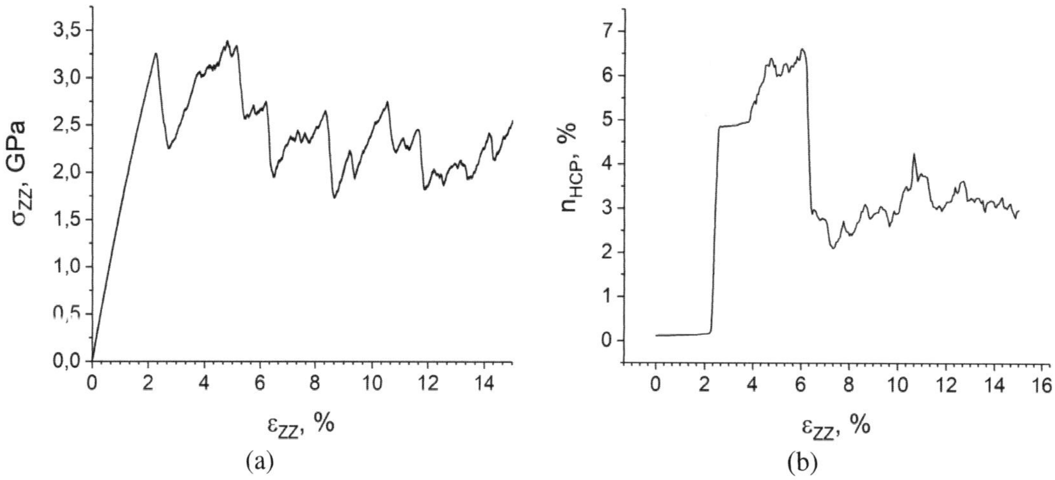

Figure 2. Dependences of stress (a) and fraction of atoms with HCP symmetry of the nearest environment (b) on the strain value.

Due to complex and interrelated deformation processes occurring in these grains, the curve in figure 2a is characterized by numerous kinks and abrupt fall in this strain interval. Starting from about 7.0% strain, the plastic deformation of the sample reaches the smallest grains. SFs are also formed in them. At about 6.3% strain, ideal structure is restored in the largest grain, while the SFs remain only in medium and small grains (figure 3c). At further straining, the largest grain is again involved in the plasticity process: full dislocations are nucleated on its boundaries and glide through the grain (figure 3d). Thus, the development of plastic deformation in samples with a gradient grain structure is characterized by pronounced stages. These stages represent a certain sequence of including grains of different sizes in the accommodation process of the internal structure for different strain intervals. Structural defects appear in the smallest grains only when larger grains have exhausted the possibilities of stress relaxation through structural rearrangements. Up to high strain values, the process of development of plastic deformation in the sample has a dislocation nature. At high strain values (more than 7.0%), structural defects nucleate and move in all grains, and deformation twins begin to grow in medium-sized grains. It should be noted that the nucleation of structural defects always occurs in the regions with the greatest deviation of the atomic volume from the equilibrium value. At the front of the propagation of the SFs the rearrangement of the structure is also realized with a local change in the atomic volume. This is in good agreement with the results of the other computer simulations [19,20].

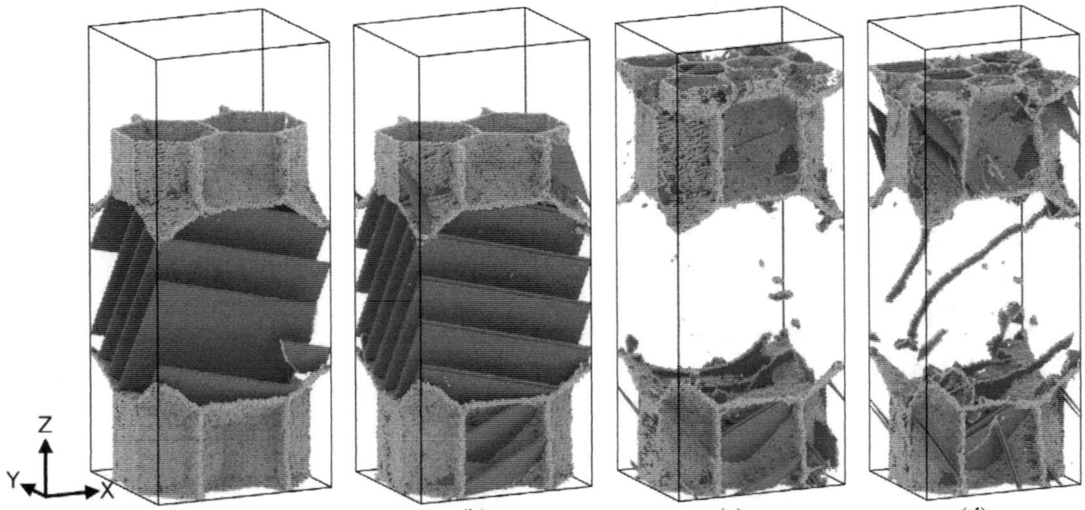

Figure 3. Fragment of the structure of the sample stretched by: 2.5% (a); 5.0% (b); 7.5% (c); 12.5% (d). Atoms with FCC symmetry of the nearest environment are not shown. Atoms with HCP, BCC and undefined symmetry of the nearest environment are marked in red, blue and gray, respectively.

Acknowledgment
The work was carried out with the financial support of Russian Science Foundation (project No. 20-79-10406).

References

[1] Ovid'ko I A, Valiev R Z and Zhu Y T 2018 *Prog. Mater. Sci.* **94** 462
[2] Ma E and Zhu T 2017 *Mater. Today* **20** 323
[3] Lu K 2016 *Nat. Rev. Mater.* **1** 16019
[4] Lu K 2014 *Science* **345** 1455
[5] Li J, Li L, Jiang C, Fang Q, Liu F, Liu Y and Liaw P K 2020 *J. Mater. Sci. Technol.* **57** 85
[6] Pan Q and Lu L 2020 *Scr. Mater.* **187** 301
[7] Guo W, Pei Z, Sang X, Poplawsky J D, Bruschi S, Qu J, Raabe D and Bei H 2019 *Acta Mater.* **170** 176
[8] Fang T H, Li W L, Tao N R and Lu K 2011 *Science* **331** 1587
[9] Yang X, Ma X, Moering J, Zhou H, Wang W, Gong Y, Tao J, Zhu Y and Zhu X 2015 *Mater. Sci. Eng.* A **645** 280
[10] Wu X, Jiang P, Chen L, Yuan F and Zhu Y T 2014 *Proc. Natl. Acad. Sci. U.S.A.* **111** 7197
[11] Chen W, You Z S, Tao N R, Jin Z H and Lu L 2017 *Acta Mater.* **125** 255
[12] Zolnikov K P, Korchuganov A V, Kryzhevich D S, Chernov V M and Psakhie S G 2019 *Phys. Mesomech.* **22** 355
[13] Zolnikov K P, Kryzhevich D S and Korchuganov A V. 2019 *Russ. Phys. J.* **62** 46
[14] Zolnikov K P, Kryzhevich D S and Korchuganov A V 2020 *Russ. Phys. J.* **63** 947
[15] Plimpton S 1995 *J. Comput. Phys.* **117** 1
[16] Zhou X W, Foster M E and Sills R B 2018 *J. Comput. Chem.* **39** 2420
[17] Honeycutt J D and Andersen H C 1987 *J. Phys. Chem.* **91** 4950
[18] Stukowski A 2010 *Model. Simul. Mater. Sci. Eng.* **18** 015012
[19] Psakhie S G, Zolnikov K P, Kryzhevich D S and Korchuganov A V 2019 *Sci. Rep.* **9** 3867
[20] Korchuganov A V, Zolnikov K P and Kryzhevich D S 2019 *Mater. Lett.* **252** 194

Comparative Analysis of Mechanical Properties of Laminated Carbon Fiber Reinforced Composites Based on PEEK and PI Matrices

P V Kosmachev, V O Alexenko and S V Panin

Institute of Strength Physics and Materials Sciences SB RAS, 2/4, Akademicheskii Pr., 634055 Tomsk, Russia

E-mail: kosmachev@ispms.ru

Abstract. The mechanical properties of laminated carbon fiber reinforced composites based on PEEK and PI matrices have been studied. Various schemes of carbon fiber lay-outing were employed. Flexural stress-strain curves, optical images of fracture surfaces and SEM micrographs of composite structure were analyzed. It has been shown that laminated composites reinforced with continuous unidirectional fibers according to layout scheme [0°/0°] possessed the maximum values of mechanical properties (flexural modulus and bending strength) under three-point bending. PEEK based composites exhibited 2 times higher flexural strength in contrast to PI based ones (0.4 GPa and 0.2 GPa, respectively), while the modulus of elasticity in bending was 4 times larger (61 GPa and 15 GPa, respectively).

1. Introduction

Research and development of polymer-based structural composites are critical trends in advanced industries worldwide. This topic is most relevant for the aerospace industry, which requires high specific strength values (the ratio of material strength to density) of the materials used. Currently, R&D is being actively carried out on the reinforcement of thermoplastic polymers reinforced with continuous fibers, which makes it possible to substantially increase mechanical properties of the composites [1-2].

The loading of polymer matrix with short fibers provides an increase in strength properties compared to neat polymer, as well as the possibility of processing by technologies traditional for most thermoplastics, i.e. extrusion and injection molding. However, there is a noticeable decrease in plasticity and corresponding embrittlement of the material, which severely limits the use of composites with short fibers. It was shown in [4] that modulus of elasticity of polymer composites increases sharply with enlargement of the length of carbon fibers.

Loading with chopped fibers (from a millimeter to tens of millimeters length) is an effective way to improve tribological and mechanical properties of composites based on polyetheretherketone (PEEK) [5-7]. The reinforcement with continuous fibers can significantly improve the performance of composites based on thermoplastic matrices. Among industrially employed continuous fibers are carbon, glass, basalt, metal ones. They have the shape of ribbons, tows or fabric, etc. [8]. The technology for producing parts of thermoplastic composites reinforced with continuous fibers consists in layering a binder (in the form of a polymer powder or a film) and fibers. For example, in [9] the properties of a layered composite consisting of 9 layers of a PEEK film (1000-300G) with a thickness of 300 μm (produced by the Victrex) and 8 layers of 3K-T300-plain carbon fiber (produced bt the Toray Industries) were studied. The stacking scheme was [0°/90°]. The fabricated composite had the following properties: ultimate bending strength of 400 MPa, flexural modulus of 20 GPa, and interlayer shear strength of 50 MPa. The highest mechanical properties of polyimide (PI) based composites were also achieved through a continuous fiber reinforcement.

The different arrangement of fibers gives rise to characteristic changes in mechanical properties, as well as to failure pattern of composite materials [10]. Mechanical properties of carbon composites with various modes of weaving and laying were studied in [11]. It has been shown that mechanical properties of the cross-reinforced composite [0°/90°/45°/-45°] are 30-40% lower compared to unidirectional composite [0°/0°]. However, authors of paper [12] noted that the anisotropy of fiberglass composites was evident when the weaving and stacking mode was [0°/0°], while the composite [0°/90°/45°/-45°] possessed isotropic properties.

The aim of this paper was to study the mechanical properties (in particular flexural modulus and bending strength) of laminated polymer composites based on polyetheretherketone and polyimide matrix, reinforced with continuous carbon fibers at variation of layout schemes.

2. Materials and methods

PEEK powder (Victrex, PEEK 450 PF, UK) with an average particle size of 50 μm and PI powder (Solver, PI-Powder, China) with an average particle size 16 μm were employed. Unidirectional carbon fabric (tape FibArm 12K-300-230, UMATEX, Russia) with areal density 230 g/m^2 and tensile strength above 4.9 GPa as well as biaxial carbon fabric (0/90-50K-1270-106, HC "Composite", Russia) with surface density 106 g/m^2 were used as a reinforcing filler.

The ratio of the components of the fabricated polymer composites (40 vol. % fiber / 60 vol. % matrix) was selected according to literature data, as the most effective for increasing mechanical characteristics.

Various layout schemes were employed (Figure 1).

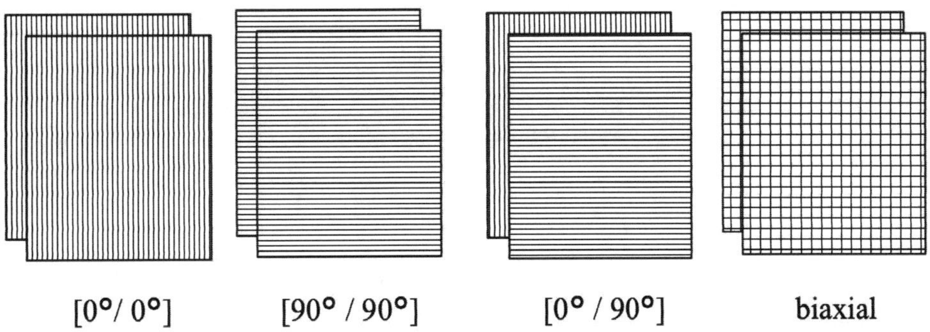

Figure 1. Layout schemes for carbon fiber layers (polymer powder layers are located between them).

After placing the layered components in the mold, the composite samples were prepared by hot pressing with the use of GT-7014-A hydraulic press (GOTECH Testing Machines Inc., Taiwan) at specific pressure of 10 MPa and a temperature of 400 °C. Cooling rate was 2 °C/min.

Shore D hardness was measured with the help of Instron 902 series machine (Instron, USA) in accordance with ASTM D2240. Mechanical properties of the composites were determined under three-point bending on an electromechanical testing machine Instron 5582 series (Instron, USA) in accordance with ASTM D790-10. Stripe shapes samples had the size of 70×10×3.5 mm. Optical microscope Neophot 2 (Carl Zeiss, Germany) was used to examine the cross-section of fractured samples (after 3-point bending). Investigations of polymer composite structures were carried out in accordance with STO TSU 041-2009 "Technique for studying the structure of solid surface by scanning electron microscopy" using a system with electron (focused) beams Quanta 200 3D (FEI Company, USA); accelerating voltage was 20 kV. Copper film was preliminary deposited onto fracture surface of composite samples.

3. Results and discussion

Figure 2 shows the diagrams of flexural stress versus relative flexural strain. Neat PEEK and PI had comparable mechanical properties (curves 1, 6). However, they decreased sharply when loaded with fibers (curves 3, 8) as well as at cross-stacking of reinforcing layers (4, 9). Biaxial reinforced composites (curves 5, 10) exhibited mechanical properties 15-20% lower compared to unidirectional reinforcement. Composites with unidirectional reinforcement had the highest mechanical properties and largest deformation when loaded transverse to the fiber orientation. The composite PEEK/C [0°/0°] (2) had a transverse flexural modulus of 61.7 GPa, while in composites PI/C [0°/0°] (7) it was equal to 15.6 GPa.

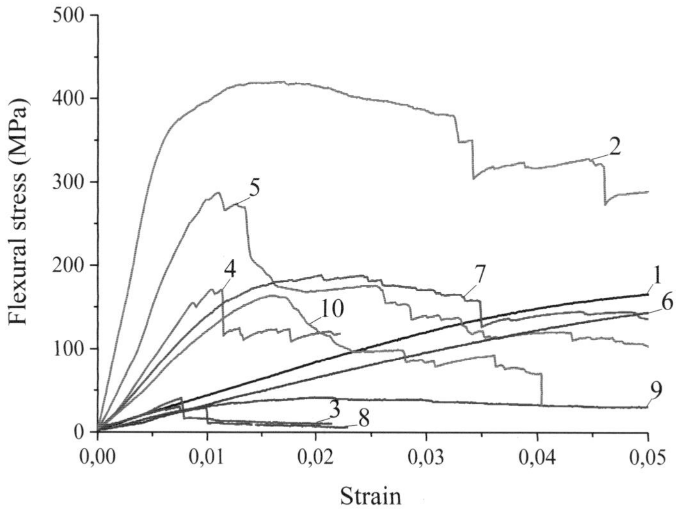

Figure 2. Flexural stress-strain curves of PEEK (1); PEEK/C [0°/0°] (2); PEEK/C [90°/90°] (3); PEEK/C [0°/90°] (4); PEEK/C [biaxial] (5); PI (6); PI/C [0°/0°] (7); PI/C [90°/90°] (8); PI/C [0°/90°] (9); PI/C [biaxial] (10).

Detailed data on physical and mechanical properties of all tested samples are presented in Table 1. It is seen that the largest increase in mechanical properties (flexural modulus and bending strength at three-point bending) was observed in composite samples # 2 and # 7, reinforced according to the [0°/0°] scheme.

Optical images of fractured sample cross section tested at 3-point bending are shown in Figure 3. Optical images of neat PEEK and PI samples are not presented since no cracking happened there.

Table 1. Physical-mechanical properties of carbon reinforced composites and unreinforced polymers.

	Sample	Density, g/cm^3	Shore D hardness	Flexural modulus G_B^U, GPa	Bending strength σ_B^U, MPa	Strain ε_{max}
1	PEEK	1.30	80.1±1.7	4.2±0.6	166.2±6.1	0.050±0.005
2	PEEK/C [0°/0°]	1.54	81.5±0.6	61.7±1.0	422.8±12.5	0.050±0.001
3	PEEK/C [90°/90°]	1.51	81.4±0.6	4.3±0.2	42.3±6.4	0.015±0.004
4	PEEK/C [0°/90°]	1.46	80.4±0.8	21.0±3.4	173.8±26.8	0.022±0.001
5	PEEK/C [biaxial]	1.33	82.0±0.7	41.2±1.8	277.2±34.9	0.050±0.001
6	PI	1.37	80.3±0.5	4.01±0.5	143.3±6.2	0.050±0.001
7	PI/C [0°/0°]	1.40	76.8±1.5	15.6±0.9	190.2±18.7	0.050±0.001
8	PI/C [90°/90°]	1.43	82.4±1.3	4.2±0.3	30.1±4.1	0.023±0.001
9	PI/C [0°/90°]	1.33	70.2±2.9	4.3±0.5	43.0±5.2	0.050±0.001
10	PI/C [biaxial]	1.38	78.6±1.2	11.9±1.3	158.4±1.2	0.040±0.001

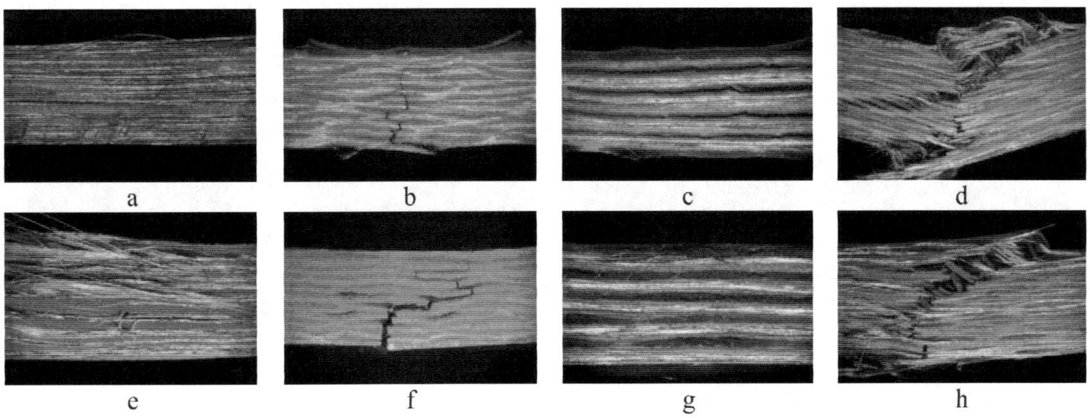

Figure 3. Optical images of cross section view of 3-point bending fractured samples: PEEK/C [0°/0°] (a); PEEK/C [90°/90°] (b); PEEK/C [0°/90°] (c); PEEK/C [biaxial] (d); PI/C [0°/0°] (e); PI/C [90°/90°] (g); PI/C [0°/90°] (g); PI/C [biaxial] (h).

It was found that cracking was slightly exhibited when sample fabrication folowed [0°/0°] layout (figure 3, a, b). Longitudinal delamination occured, while interlayer cracks took ~1/5 of the sample thickness. Through transverse cracking occurred when lay-outing out was [90°/90°] (figure 3, b, f). Tensile, shear and delaminating components of cracking were evident. Pronounced interlayer delamination took place when the [0°/90°] scheme (c, g) was employed. Exhibited shear cracking were characteristic when the [biaxial] scheme (d, h) was employed.

SEM micrographs of composite structures are shown in Figure 4. One can see that regions with incomplete wetting of carbon fibers presented in the composites. Thus, in order to further increase strength properties the problem of matrix material better penetration into fiber layers should be solved.

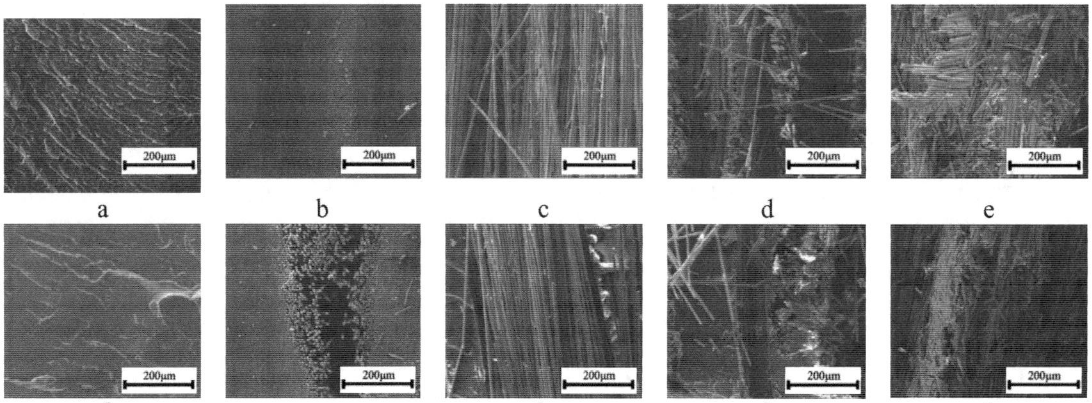

Figure 4. SEM micrographs of composite structures: PEEK(a), PEEK/C [0°/0°] (b); PEEK/C [90°/90°] (c); PEEK/C [0°/90°] (d); PEEK/C [biaxial] (e); PI (f); PI/C [0°/0°] (g); PI/C [90°/90°] (h); PI/C [0°/90°] (i); PI/C [biaxial] (j).

4. Conclusions

It has been shown that laminated composites reinforced with continuous unidirectional fibers according to layout scheme [0°/0°] possessed the maximum mechanical properties (flexural modulus and bending strength) under three-point bending. PEEK based composites possessed flexural strength 2 times higher compared to PI based ones (0.4 GPa and 0.2 GPa, respectively), while the modulus of elasticity in bending is 4 times larger (61 GPa and 15 GPa, respectively).

According to the obtained results the use of PI powder for fabricating laminated carbon fiber reinforced composites is of low prospect because of low mechanical properties (in comparison with PEEK based composites). The most probable reason is the lower melt flow index of the PI matrix.

Acknowledgments

The reported study was funded by RFBR, project number 19-38-60064\19 and the Government research assignment for ISPMS SB RAS, Project No. III.23.1.3.

References

[1] Hassan A M et al. 2018 Comp. Part A: Appl. Sc. and Manuf. **112** 155–160
[2] Wang X et al 2020 Appl. Surf. Sc. **510** 145400
[3] Garcia-Gonzalez D et al 2015 Comp. Struct. **133** 1116
[4] Panin S V et al 2020 Materials **13** 338
[5] Lin L et al 2019 Wear **432–433** 202928
[6] Chu X X et al 2010 Cryogenics **50** 84
[7] Zhao X et al 2017 J Bionic Eng **14** 640
[8] Kerber M L et al 2008 Polymer composite materials: structure, properties, technologies *Saint-Petersburg: Professiya*
[9] Hassan E A M et al 2018 Comp. Part A: Apppl. Sc. and Manuf. **112** 155
[10] Dhand V et al 2015 Comp.Part B: Eng. **73** 166
[11] Morioka K and Tomita Y 2000 Mater. Charact. **45** 125
[12] Zhao W-S and Ling L 2010 Plast Addit. **79 (1)** 45–48

Some approaches for residual porosity estimating

D Yu Kosyanov[1], A P Zavjalov[1,2], A A Vornovskikh[1], A M Zakharenko[1], X Liu[3,4] and J Li[3,4]

[1]Far Eastern Federal University, 10 Ajax Bay, Russky Island, Vladivostok 690922, Russian Federation
[2]Laboratory of Methods of Synchrotron Radiation, Institute of Solid State Chemistry and Mechanochemistry, Siberian Branch of the Russian Academy of Sciences, 18 Kutateladze Street, Novosibirsk 630128, Russian Federation
[3]Key Laboratory of Transparent Opto-functional Inorganic Materials, Shanghai Institute of Ceramics, Chinese Academy of Sciences, 585 He-shuo Road, Shanghai 201899, China
[4]Center of Materials Science and Optoelectronics Engineering, University of Chinese Academy of Sciences, Beijing 100049, China

E-mail: vornovskikh.aa@gmail.com

Abstract. Several approaches for materials residual porosity assessment are presented in this paper. Importance of development of *accurate* model is shown based on Al_2O_3–Ce:YAG and Al_2O_3–Ce:(Y,Gd)AG ceramic phosphors taken as model objects. The evaluated difference in the outcomes of different methods applied for estimating porosity has been shown to be as high as ~40%. The obtained results become applicable to another types of inclusions within the studied volume of a material as in the models under consideration the main role is given to a geometrical perspective of the problem.

1. Introduction

Evaluation of the material's intrinsic porosity using data collected from the cross-section is a special case of the wide-spread problem in materials science related to determining the content of various non-uniformities. As key defects, the surfaces/slices of dense materials (single crystals, glasses, ceramics) observe pores as cross-sections and the question about their concentration and content in the material's bulk becomes important [1, 2].

The simplest way to answer it is to inscribe pores into the found cross-sections and to evaluate their volume. However, it is very unlikely that the cross-section passes exactly through the pore centers to regard such an identification as an accurate one. It has been shown in work [3] that the size of a random cross-section is $4/\pi$ times lower than the size of a spherical pore. Authors replace all found pore cross-sections with the mean pore size $<d>$, which is additionally scaled-up by a factor of $4/\pi$. However, this consideration is quite rough approximation. Pores of diameter D are shown on cross-section only in the case if the slice plane is closer or stay at D/2 from the pore center. As a result, finer pore fractions are represented on a cross-section in a smaller proportion to the content of this fraction in the bulk, than lager pores [4]. This aspect can be corrected using multiplier $1/D$ and renormalization of the size distribution obtained during the material study. The more significant issue, when pores of D size can give different cross-sections d, is impossible to take into account through simple reasoning.

Therefore the estimations of pore volume fraction θ turn out to be rough in different versions of such constructions.

2. Theoretical section

In the previous work [4], we performed consistent reasoning and found an expression to connect the distribution density of pore cross-sections sizes $f(d)$ with the pore size distribution density $g(D)$ (1):

$$f(d) = \frac{d}{L}\int_d^{+\infty} \frac{g(D)\mathrm{d}D}{\sqrt{D^2-d^2}} \tag{1}$$

The following relation was obtained at the intermediate stage of expression derivation (1):

$$nSd\left(\int_d^{D_{\max}} \frac{g(D)\mathrm{d}D}{\sqrt{D^2-d^2}}\right)\mathrm{d}d = N_S f(d)\mathrm{d}d \tag{2}$$

where S – area of the explored cross-section of material, n – a number pore concentration in the volume V. The combination nLS was used as N_S in the work [4], where L – thickness of considered layer of material. However, this is inaccurate, because expression (2) is number of slices with dimensions $[d;\ d + \mathrm{d}d]$. Although in the expression (1) L value has another meaning and plays the role of a normalizing factor. The exact meaning of value N_S in expression (2) is the total number of pores on cross-section. It is also shown in the intermediate derivations that the combination of values $(D/L)nLSg(D)\mathrm{d}D = nDSg(D)\mathrm{d}D$ expresses the pores number of sizes $[D;\ D + \mathrm{d}D]$, which occurred to be sliced [4]. Further integration leads to the expression $N_S = nS\langle D\rangle$. Thus, the meaning of the normalizing factor L was clarified: the mean pore size in the bulk of material $\langle D\rangle$.

In framework of the *accurate* model number concentration of pores n and their volume concentration θ can be calculated using data from studied slice of the material without necessity of direct calculation of function $g(D)$. The corresponding expressions (3, 4) can be obtained from (1) by inverting the integral equation and direct calculating the required values $\langle D\rangle$ and $\langle D^3\rangle$:

$$n = \frac{N_V}{V} = \frac{N_S}{S\langle D\rangle} = \frac{2}{\pi}\langle\frac{1}{d}\rangle\frac{N_S}{S} \tag{3}$$

$$\theta = \frac{V_p}{V} = \frac{\pi}{6}n\langle D^3\rangle = \frac{\pi}{6}\frac{N_S}{S}\frac{\langle D^3\rangle}{\langle D\rangle} = \frac{N_S}{S}\frac{\pi\langle d^2\rangle}{4} = \frac{S_p}{S} \tag{4}$$

for expressions (3-4): V – volume of material; N_V – the total number of pores in volume V; S – area of the explored cross-section of material; N_S – the total number of pores detected in the cross-section S; V_p – total pore volume in the material's bulk V; S_p – total surface area of pore sections on the studied slice of the material; the brackets <…> denote the average value of distribution.

Let's consider another two models for estimating porosity. The first of them, the *simplest* model, assumes inscribing pores of the same size into the found cross-sections. The derivation of expression (1) contains a multiplier (D/L). It appears because only fraction of pores in the considered layer of thickness L occurs to be sliced – pores in the layer of less thickness D. Thus, reverse multiplier should be applied to determine pore size distribution in the materials bulk. In the language of distribution function, this means that $g(D) = (L/D)f(D)$, where L simply plays the role of the normalizing coefficient and equal to the harmonic mean $\langle 1/d\rangle^{-1}$. For such model we obtain expressions (5, 6):

$$n = \frac{N_S}{SL} = \frac{N_S}{S}\langle\frac{1}{d}\rangle \tag{5}$$

$$\theta = \frac{V_p}{V} = \frac{\pi}{6}n\int\frac{\langle 1/d\rangle^{-1}}{d}f(d)d^3dd = \frac{N_S}{S}\frac{\pi}{6}\int f(d)d^2dd = \frac{N_S}{S}\frac{\pi\langle d^2\rangle}{6} = \frac{2}{3}\frac{S_p}{S} \tag{6}$$

The second model, *modified*, is based on work [3]. It was shown that the cross-section size of a spherical pore is $4/\pi$ times lower than its size. Therefore, we use the same approach as for the *simplest* model, but increase pore sizes by this coefficient. In the language of distribution function, we obtain the expression $g(D) = (L/D)f(\pi D/4)$, that adds $4/\pi$ multiplier of appropriate power in the expressions for calculating the average values of n and θ (7, 8):

$$n = \frac{N_S}{SL} = \frac{\pi}{4}\frac{N_S}{S}\langle\frac{1}{d}\rangle \tag{7}$$

$$\theta = \frac{V_p}{V} = \frac{N_S}{S}\left(\frac{4}{\pi}\right)^2\frac{\pi\langle d^2\rangle}{6} = \frac{32}{3\pi^2}\frac{S_p}{S} \approx 1.08\frac{S_p}{S} \tag{8}$$

3. Experimental section

Validation and comparative analysis of presented approaches for assessing the residual porosity of materials was conducted using model – Al_2O_3–Ce:YAG and Al_2O_3–Ce:(Y,Gd)AG composite ceramic phosphors. A series of composite ceramics were prepared by solid-state reactive sintering in vacuum using high purity commercial oxides. In brief, the molar ratio of Al_2O_3/YAG was controlled to be 0.75; the Ce^{3+} and Gd^{3+} concentrations were set to be 0.1 at% and 25 at%, respectively, according to the stoichiometric ratio of (Ce+Y+Gd):Al=3:5 for YAG phase. 0.8 wt% tetraethyl orthosilicate (TEOS) and 0.08 wt% MgO were used as sintering aids. Powder mixtures were ball milled, dried, sieved, and air-annealed. The green bodies were cold isostatic pressed, and sintered at 1765°C under vacuum of 5.0×10^{-4} Pa for 4, 8 and 12 hours. Finally, composite ceramics were air-annealed to recover oxygen stoichiometry and mirror polished for further measurements.

The porosity of ceramic samples was characterized by confocal laser scanning microscopy (CLSM, LSM 800, Zeiss, Germany) using a laser with wavelength of 642 nm and 63x zoom. After acquiring elementary sample sections ("slices") of about 100×100 μm^2, which were separated to each other by 250 nm in depth, the 3D volume ($100\times100\times15$ μm^3) was rebuilt by using appropriate software (Arivis AG). To calculate the section pore size and to quantify the number of the pores from the cross-section, all slices were merged along z axis and were represented in the xy plane. The residual porosity was calculated by *simplest*, *accurate*, and *modified* (using Ref. [3]) models.

4. Results and Discussion

Pore section size distribution and the residual porosity of model ceramic samples synthesized at 1765°C for 4, 8 and 12 hours are presented on Figure. 1a,b. Some reduction of mean pore sizes and their size variance with the increase of sintering duration from 4 to 12 hours is caused by completion of densification at higher densities. Samples co-doped with Ce^{3+} and Gd^{3+} in the range of 4-8 hours sintering features lower driving forces of densification due to incomplete formation of Ce:(Y,Gd)AG solid solution. As a result, within this time interval the mean pore section size increases by 13% till 223 nm with the porosity being reduced by 19% down to 0.0076 vol.% (Figure. 1a, inset; 1b). Anyway, after 12-hour isothermal treatment, Al_2O_3–Ce:(Y,Gd)AG composite ceramics has "pore-free" structure. In turn, Al_2O_3–Ce:YAG composite ceramics possesses larger pore amount per unit volume simultaneously with their lower size variance (Figure. 1a,b).

Evaluated by the *accurate* model (4) residual porosity values for Al_2O_3–Ce:YAG and Al_2O_3–Ce:(Y,Gd)AG samples range in 0.0167-0.0065 vol.% and 0.0093-0.0049 vol.%, respectively (Figure. 1b). Taking into account equal concentrations of the used SiO_2 and MgO additives, observed values prove some positive influence of Gd^{3+} ions on densification of Ce:(Y,Gd)AG ceramic solid solutions on the final stage of sintering.

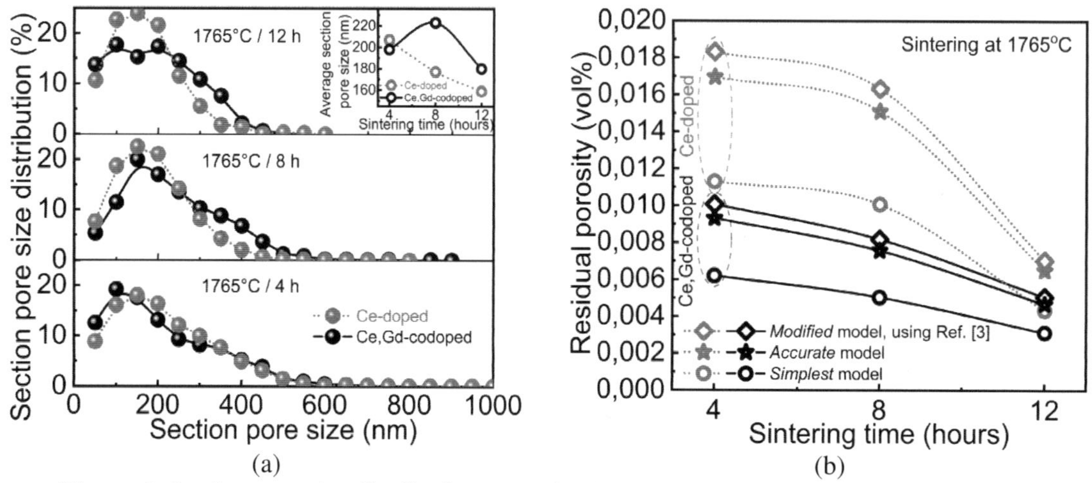

Figure 1. Section pore size distribution (a) and residual porosity (b) of Al_2O_3–Ce:YAG and Al_2O_3–Ce:(Y,Gd)AG composite ceramic phosphors obtained by reactive sintering in vacuum at 1765°C for 4-12 hours.

Results of the *simplest* model (6) are underestimated by ~33%, the ones obtained by the *modified* model (8) based on [3] are overestimated by ~8%. Thus, the total difference of porosity estimations reaches ~40%. As a result, the lower limit of the porosity estimation for Al_2O_3–Ce:YAG ceramics is comparable to the upper limit of the porosity estimation for Al_2O_3–Ce:(Y,Gd)AG ceramics (Figure. 1b). In practice, this can lead to wrong conclusions about material properties when comparing data obtained using different porosity estimation models. It is a bright example of the necessity to clearly understand, which model for porosity assessment was used in a certain literary source, technical documentation, etc.

5. Conclusions

Based on the integral connection between pore size distribution in the bulk and pore section size distribution observed on the slice, *accurat*e model is demonstrated to establish the "intuitive" method for porosity assessment as a fraction occupied by the pore sections on the studied slice of the material. Comparison with another two models was carried out. The evaluated difference in the outcomes of different methods applied has been shown to be as high as ~40% based on Al_2O_3–Ce:YAG and Al_2O_3–Ce:(Y,Gd)AG composite ceramics taken as model objects. Such a huge variance among the calculated values demonstrates the importance of the analysis of each certain porosity assessment approach applied when comparing the experimental data.

Acknowledgments

This work was supported by the Russian Science Foundation (Project No. 20-73-10242).

References

[1] Rice R W 1995 The porosity dependence of physical properties of materials: A summary review *Key Engineering Materials* **115** 1-20

[2] Messing G L and Stevenson A J 2008 Materials science: Toward pore-free ceramics *Science* **322** (5900) 383-4

[3] Zhang W, Lu T, Wei N, Wang Y, Ma B, Li F, Lu Z, Qi J 2012 Assessment of light scattering by pores in Nd:YAG transparent ceramics *J. Alloys Compd.* **520** 36–41

[4] Kosyanov D Yu , Yavetskiy R P, Parkhomenko S V, Doroshenko A G, Vorona I O, Zavjalov A P, Zakharenko A M, Vornovskikh A A 2019 A new method for calculating the residual porosity of transparent materials *J. Alloys Compd.* **781** 892–7

IOP Conf. Series: Materials Science and Engineering 1093 (2021) 012016 doi:10.1088/1757-899X/1093/1/012016

Development of methodological approaches for non-destructive testing of welded joints of steel finned panels by the active thermography method

I O Kotovshchikov[1,2], Ju O Yakovlev[3,4] and V E Prohorovich[1]

[1]Department of Control Systems and Robotics, ITMO University, Saint-Petersburg, Russia
[2] Locus Ltd., Saint-Petersburg, Russia
[3]Department A8 "Engines and power plants of aircrafts", BSTU «VOENMEH» named after D.F. Ustinov, Saint-Petersburg, Russia
[4]Scientific and technical center "Etalon", Saint-Petersburg, Russia

E-mail: kotovshchikov.ilya@mail.ru

Abstract. This paper examines methodical approaches to non-destructive testing (NDT) of thin steel T-welded joints made by ultrasound welding. The choice of the active thermography method is justified. Numerical parameters were developed for characterize the operational properties of the plate-fin heat exchanger and the results of finned panel inspection by the active thermography method. These parameters were compared and the correlation was found. The developed methodological approaches will be useful in the future for developing an NDT technique that allows detecting internal defects in welded seams which decrease breaking load and worsen the operational properties of products made by finned panels.

1. Introduction

The relevance of the developing methodological approaches to NDT of welded seams of steel finned panels is due to the peculiarities of the production technology of finned panels and products made from them. Finned panels are used in systems with separation of gas and liquid flows (plate-fin heat exchangers), which means that through defects are strictly not allowed. Practice has shown that through defects can occur in a product during operation process.

The manufacturing technology of the plate-fin heat exchangers from finned panels consists of the following main stages:

1. Thin steel sheets (thickness about 1.5 mm) are wound on a special reel.
2. The automatic welding machine feeds the perpendicular steel sheet, which will be the fin of the panel.
3. The fins are unilaterally welded to the panel by the ultrasonic welding during the reel rotation process (type T1 of the welded seams according to GOST 5264-80).
4. The finned panel is removed from the drum and unfolded.
5. Several finned panels are stacked on top of each other to form the body of the final product.
6. Finned panels are welded to each other by manual arc welding.

7. Carrying out NDT of welded joints obtained at stage 6.
8. The finalizing of the product.

To date, quality control of welded joints of finned panels is carried out by penetrant test method and only after the end of the manufacture of the heat-exchange sub product (after stage 6). Through defects with a characteristic size of from 0.1 mm, arising after manual welding, are reliably detected by this method. This method has such disadvantages as low control productivity, the necessity of work with toxic substances, and the impossibility of detecting internal defects.

A violation of the conditions of the welding environment, imperceptible to the operator, during the process of automated ultrasonic welding, can lead to the occurrence of internal defects such as lack of fusion. Technological stage 4 leads to the occurrence of mechanical stress in the welded seams, which can cause the occurrence of internal defects. Thereby, an internal defect in the T-welded seams of finned panels, that reduce the strength of the welded joint, may appear during stage 3 and stage 4 of the production process. Further technological operations cannot lead to the occurrence of internal defects, and after welding the panels to each other (stage 6), the seams made by ultrasound welding will become inaccessible for NDT. Thus, the checking for internal defects in finned panels after their unfolding (stage 4) is technologically justified.

Internal defects of finned panels are not detected by NDT tools used today. To identify them, it is proposed to use the thermal method of inspection - active thermography. This paper describes the development of the main methodological approaches to the creation of a technique for NDT of T-shaped welded joints of thin-walled plates of plate-fin heat exchangers obtained by ultrasonic welding.

Due to the peculiarities of the production technology of finned panels and operation conditions of the product, the active thermography method was chosen to search for internal defects. This NDT method use thermal excitation of the part surface and registration of the dynamics of its thermal response using an infrared (IR) camera [1-5]. Four halogen lamps were used for thermal excitation in this research. The active thermography system registers the dynamics of the controlled object surface temperature in the form of a sequence of IR images (thermograms). Further, using special software, the obtained information about the temperature dynamics is processed by software algorithms [6].

The thermal response of the controlled object surface is determined by such factors as the thermophysical properties of the material, the thickness of the object, peculiarities of the internal structure, the presence of internal defects, etc. One of the effective approaches for flaw detection by the active thermography method is to compare the obtained thermograms with the reference ones obtained for defect-free samples under the same conditions.

The purpose of this research is to develop methodological approaches to the control of welded joints of steel finned panels for the presence of internal defects. To achieve this goal, the following tasks were formulated:

1. To identify the features of displaying a defective welded joint on a thermogram.
2. To reveal the parameter numerically characterizing the presence of unacceptable internal defects on the thermograms of the welded joints.
3. To reveal the correlation between the operational properties of the products and thermal response of the welded joints on the thermograms.

2. Research

For the research, samples were prepared, which are fragments of finned panels, shown in Figure 1. In order to identify the peculiarities of the imaging of a defective welded joint on the thermogram, all samples were tested using an active thermography system, from the flat side. The presence of fins on the reverse side of the samples leads to heat removal in the zone of welded seams. Thus, seams should appear on the thermogram as darker areas relative to the surrounding area. If the seam contains internal defects such as lack of fusion, then in the zone of defects, the thermal conductivity process will be disrupted, and the defective seam will have an inhomogeneous thermal image on the thermogram.

The samples are fragments of two finned panels. Each sample is numbered in accordance with its belonging to one of the two panels and the order in which it follows (Figure 1).

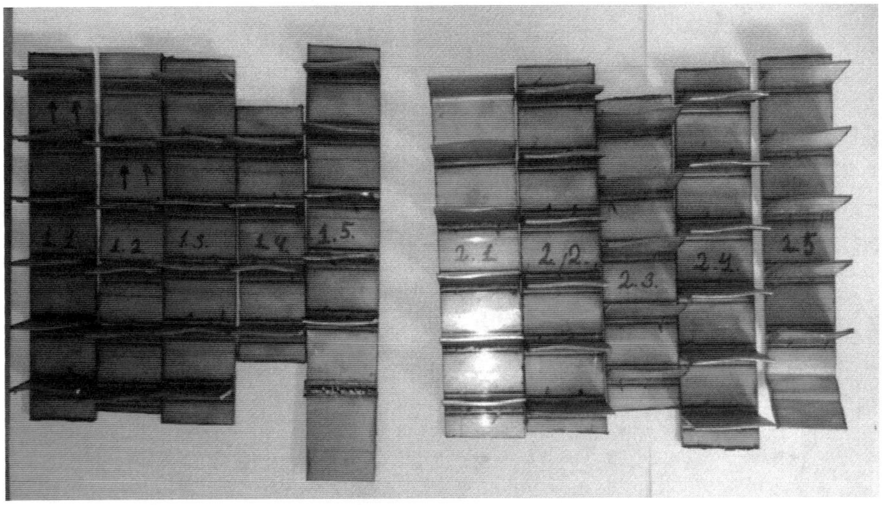

Figure 1. General view of the researched samples (fragments of finned panels).

2.1. Thermographic research of samples

The features of the researched welded joints cause to the following features of their thermograms: edge effect, reflections of foreign objects on the metal surface of the controlled object, the influence of surface inhomogeneities (such as scratches) on the thermal response, and inhomogeneity of heating along the seams. Thus, the NDT technique of finned panels using active thermography method should include the following approaches: matting the surface with black paint to eliminate glares and the influence of scratches; do not evaluate the defectiveness of welded seams areas located close to the edges of the panel.

Figure 2 shows thermographic images of the studied samples, which are fragments of finned panels. Seams 1.2.2 and 1.3.6 have an inhomogeneous thermal response, which can be caused by internal defects such as lack of fusion.

Thus, a qualitative parameter for defectiveness can be considered the brightness difference (inhomogeneous thermal response) along the welded joint, as in the areas highlighted by ellipses in Figure 2.

The parameter that numerically characterizes the presence of internal defects such as lack of fusion in the thermograms of the welded joints can be the "delta" parameter:

$$\Delta = I_{max} - I_{min} \text{, where}$$

I_{max} is the maximum intensity of IR radiation along the section of the welded seam, and I_{min} is the minimum intensity. Both intensity values should take into account the edge effect: at the edges of the fragments, the intensity of the thermal response is overestimated.

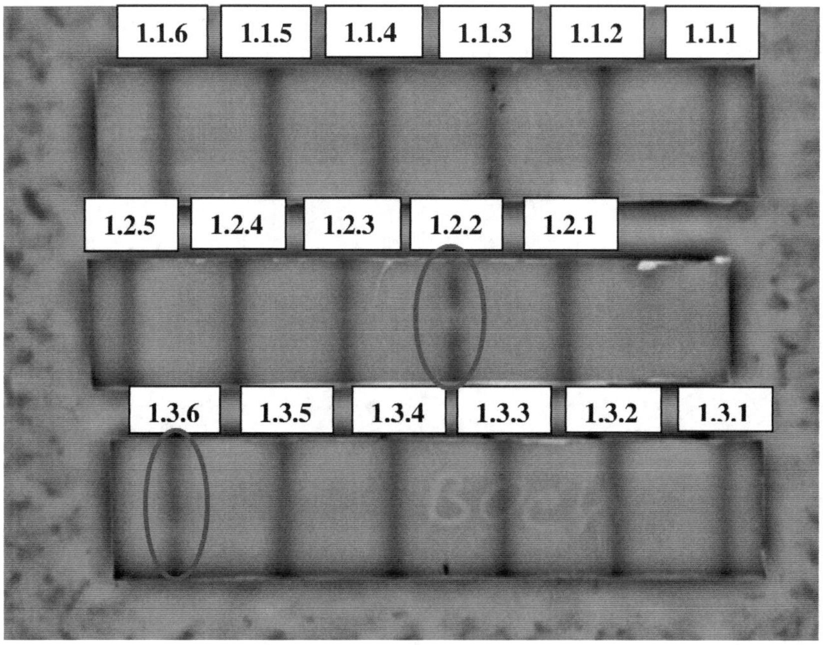

Figure 2. Thermograms of samples 1.1-1.3. Ellipses mark seams with inhomogeneous thermal response that can be defects such as lack of fusion.

2.2. Evaluation of the breaking load of the samples' seams

One of the key performance characteristics of the welded joints of the finned heat exchanger panels is the breaking load. By exerting pressure on a poorly welded wall, gas or liquid can destroy the T-welded joint. The load, after which the fracture of the welded joint occurs, corresponds to the maximum allowable stress of the product, called the breaking load. The higher the breaking load of the welded seams of the fragments, which characterizes the strength of the finned panel, the higher the reliability of the finished product and longer its life cycle. Therefore the breaking load can numerically describe the performance characteristics of the products. To reveal the correlation between the breaking load of the researched welded seams and their thermal response on the thermograms, the strength of the welded joints was evaluated by the destructive method. Before testing, each fragment was cut into pieces containing one seam equidistant from the edges of the piece. The test procedure used is described in the relevant standards [7-8].

The algorithm for evaluating the welded joints breaking load was as follows:

1. Preparation of samples-pieces of a welded joint.
2. Place the specimen in the test bench with the welded bead facing down as shown in Figure 3.
3. Vertical loading by using special tooling.
4. Registration of the graph of the load on time.
5. Determination and registration of the maximum load P_{max}, after which the sample failed (fracture occurred).

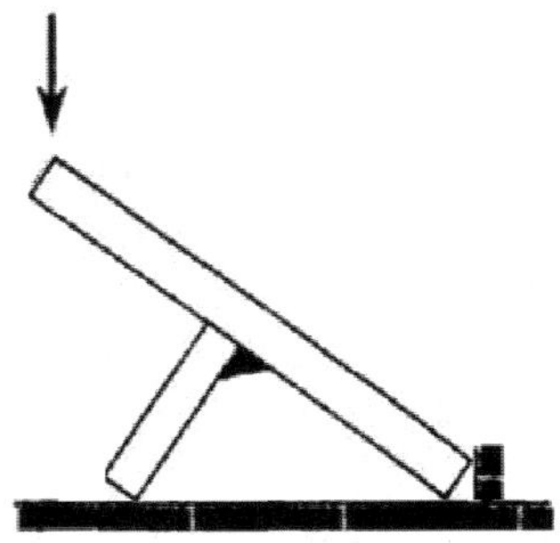

Figure 3. The layout of the sample loading *P* [8].

2.3. Comparison between the operational properties of products and thermal response of the welded joints on the thermograms

According to the results of the study, the values of the breaking load and the "delta" parameter were obtained for samples of welded joints of finned panels. To search for correlations between the values of Δ, characterizing the results of NDT by the active thermography method and the values of P_{max}, characterizing the operational properties of products, a comparative diagram was built, shown in Figure 4. It can be seen from the diagram that welded seams, with a sharp excess of the average value of the parameter "delta" (seams # 8, 15, 18 in the diagram), have significantly lower strength than seams with a homogeneous thermal response.

For defect-free seams with a breaking load value $P_{max} \geq 1.5$ kN, the average value of the parameter is $\Delta \approx 5 \pm 2$ c.u. (conventional units).

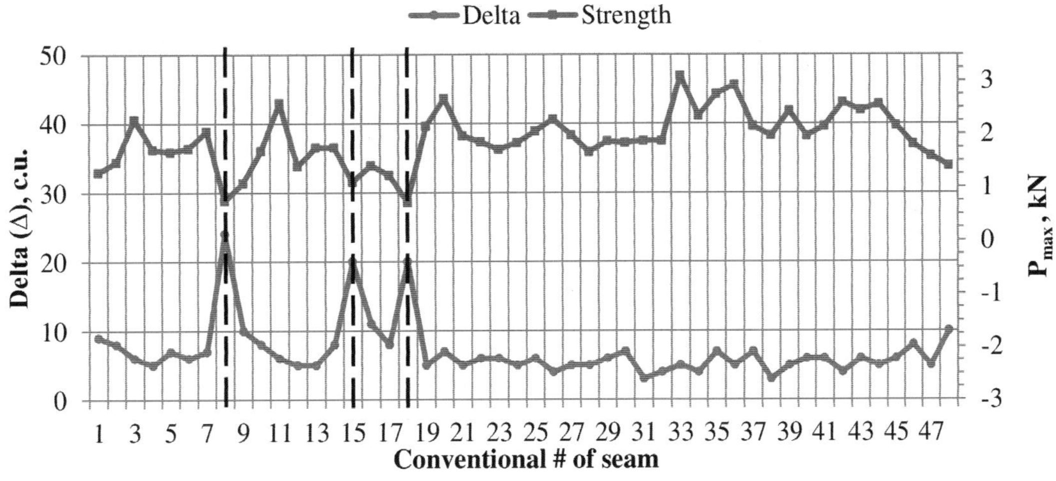

Figure 4. Comparative diagram of strength and parameter Δ of the researched seams. The dotted lines mark the points corresponding to seams 1.2.2, 1.3.6 and 1.5.1.

IOP Conf. Series: Materials Science and Engineering 1093 (2021) 012016 doi:10.1088/1757-899X/1093/1/012016

3. Conclusions and perspectives

The defectiveness of seams with a high value of the parameter Δ – welded seams 1.2.2, 1.3.6 and 1.5.1 is confirmed. An inverse dependence of the strength of the welded seam on its Δ parameter, obtained during NDT by the active thermography method, is observed: the higher the "delta" value means the lower the strength of the welded joint. In this case, the main factor that reduces the strength of the welded joint is the presence of internal defects such as lack of fusion in the seam. Therefore, the developed parameter Δ numerically characterizes the presence of internal defects that reduce the strength of the welded joint.

The P_{max} value is influenced by many factors, and not all seams with a high value of the Δ parameter failed at a load of less than 1.5 kN. However, all seams with a value of the thermal response inhomogeneity parameter $\Delta > 5 \pm 2$ c.u. had a reduced strength. Thus, according to the currently developed approaches, there is an over rejection. It should be noted that all rejected welded seams have an excess of the threshold level by 1 c.u., which is less than the measurement error of the Δ parameter.

For the further development of the NDT technique, it is necessary to increase the number of samples and carry out work on scaling the developed methodological approaches for longer welded seams.

References

[1] Ovechkin B B 2006 Basics of heat engineering. Transfer of energy and mass: study guide *TPU Publishing House*
[2] Budadin O N, Potapov A I, Kolganov V I and Troitsky-Markov T E 2002 Thermal non-destructive testing of product *Moscow: Nauka*
[3] Budadin O N, Vavilov V P and Abramova EV 2011 Thermal control *Moscow: ID Spectrum*
[4] Vavilov V P 2009 Infrared thermography and heat control *Moscow: Spektr*
[5] Kotovshchikov I O 2019 *IOP Conf. Ser.: Mater. Sci. Eng.* **683** 012076
[6] Kotovshchikov I O 2018 *Proc. Russian. Conf. "TestMat"* (Moscow) pp 119–129
[7] GOST R 53690-2009. Approval testing of welders – Fusion welding – Part 1: Steels (MOD)
[8] RD 03-495-02 Process regulations for certification of welders and welding specialists

Results of long-term permeability of small-diameter vascular prostheses modified by pro-angiogenic factors and athrombogenic drug coating (model of large laboratory animals)

E O Krivkina[1], A V Mironov[1], A R Shabaev[1], M A Rezvova[1], E A Velikanova[1], V O Tkachenko[2], M Yu Khanova[1], T Y Sergeeva[1], S S Krutitskiy[1] and L V Antonova[1]

[1]Research Institute for Complex Issues of Cardiovascular Diseases, Kemerovo, 650002, Russian Federation
[2]Budker Institute of Nuclear Physics, Novosibirsk, 630090, Russian Federation

E-mail: leonora92@mail.ru

Abstract. Developed for bypass surgery, tissue-engineered vascular grafts should remain passable after implantation and have high rate of endothelialization to ensure long-term patency, atrombogenicity and biocompatibility. This study was aimed at assessing the long-term patency of polyhydroxybutyrate/valerate (PHBV)/polycaprolactone (PCL) vascular grafts modified by pro-angiogenic factors and athrombogenic drug coating in a large laboratory animal model.

1. Introduction

Tissue-engineered small diameter vascular grafts are the alternative to the deficiency of autologous blood vessels in the surgical treatment of cardiovascular diseases. Consisting of biocompatible and biodegradable polymers, these functionally active grafts have been developed all over the world [1]. However, tissue-engineered vascular grafts must have a high patency, athrombogenicity and biocompatibility [2]. One of the ways to increase biocompatibility in polymer grafts is to incorporate growth factors (GFmix) into the tubular graft sheath and additionally modify the surface by antiplatelet agents and anticoagulants [1,3].

2. Materials and methods

2.1. Fabrication of tissue-engineered vascular grafts

Biodegradable vascular grafts (4 mm diameter) were prepared by electrospinning (Nanon-01A, MECC, Japan) from 14% (poly(ε-caprolactone) PCL and 10% (polyhydroxybutyrate/valerate) PHBV solution dissolved in chloroform, with the incorporation of bioactive molecules (GFmix). Vascular endothelial growth factor (VEGF; Sigma-Aldrich, USA) was incorporated into the inner third of the graft wall. The basic fibroblast growth factor (bFGF; Sigma-Aldrich, USA) and stromal cell–derived factor-1α (SDF-1α; Sigma-Aldrich, USA) were incorporated into the outer two thirds of the graft wall.

2.2. Surface modification of the polymer grafts by anticoagulants and antiplatelet agents

Using authors` original method [4], the inner surface of the graft was additionally modified by antiplatelet agents and anticoagulants to increase the thrombosis resistance. The antiplatelet and anticoagulant agents - iloprost and unfractionated heparin, respectively, were used as antithrombotic drugs. Using the method of radiation-induced polymerization, polyvinylpyrrolidone (PVP) hydrogel coating was formed on the inner surface of the sheath to bind heparin and iloprost. Firstly, the graft was submerged in a PVP solution in ethanol at a concentration of 1-25% for 30 minutes. Then the prosthesis was removed from the solution, air-dried for 24 hours, placed in argon-filled glass vials and irradiated at 25-50 kGy. To bind heparin and iloprost to the PVP hydrogels, vascular prostheses were kept in a modifying solution of unfractionated heparin and iloprost in a glycine buffer solution for 30 minutes. The grafts were then air-dried under sterile conditions.

2.3. In Vivo Implantation

PHBV/PCL/GFmix$^{Hep/Ilo}$ prostheses of 4 mm diameter and 40 mm length were implanted into the carotid artery of a sheep, according to the scheme: 1 animal – 1 prosthesis (n=8); resection of a section of the carotid artery of 40 mm length for the implantation of the graft using the "end-to-end" method. The implantation period is 18 months. The comparison group - Gore-Tex® vascular prostheses (Gore-Tex, USA) (n=5) were implanted into sheep carotid artery for 6 months. Ultrasound dopplerography was performed on the M7 Premium device (Mindray, China) to assess the patency of the implanted grafts: at days 1 and 5, months 1, 3, 6, 9, 12, 15 and 18 following surgery – for passable vascular prostheses; at days 1 and 5 – for thrombosed prostheses.

2.4. Histological examination

PHBV/PCL/GFmix$^{Hep/Ilo}$ grafts were explanted after 1.5 years of implantation, whereas Gore-Tex grafts were explanted after 6 months of implantation. To determine the presence of calcium, histological examination of explanted samples stained with hematoxylin-eosin, Van Gieson, orsein, alizarin red S and DAPI was performed. The explanted samples were fixed in formalin for 24 hours, rinsed with running tap water and dehydrated in IsoPrep (BioVitrum, Russia). Then they were soaked 3x in paraffin at 56 °C for 60 minutes. Impregnated samples were immersed in HISTOMIX® paraffin (BioVitrum, Russia). 8 μm sections were cut using the HM 325 microtome (Thermo Scientific, USA), placed in a thermostat and dried overnight at 37 °C. After drying, they were dewaxed 3x in o-xylene for 1-2 minutes and dehydrated 3x in 96% ethanol for 1-2 minutes. Following that, the dewaxed sections were stained in accordance with the study protocol. The sections were evaluated by light and fluorescence microscopy (AXIO Imager A1 microscope with 50x, 100x and 200x magnification, Carl Zeiss, Germany).

2.5. Immunofluorescence examination

The explanted grafts were frozen at -120 °C and fixed in a Tissue-Tek embedding medium (Sakura, Japan). 8 μm sections were cut using a CryoStar NX50 cryostat (Thermo Scientific, USA). Sections were stained with antibodies to CD31, vWF, α-actin, collagen I, III, IV type (Abcam, England), cell nuclei were contrasted with DAPI (Sigma, USA). Control samples were prepared similarly to experimental ones, but 1% BSA was used instead of primary antibodies. Slides were mounted onto a labeled glass slide with ProLong (Life Technologies) and analyzed using a laser scanning microscope LSM 700 (Zeiss, Germany).

3. Results and discussion

3.1. Results of histological assessment

A day after implantation, 100% of Gore-Tex prostheses and 37.5% of PHBV/PCL/GFmix$^{Hep/Ilo}$ grafts were thrombosed. After 10 months of implantation, another PHBV/PCL/GFmix$^{Hep/Ilo}$ graft was thrombosed. Thus, the patency of this group of prostheses after 1.5 years of implantation was 50%. Aneurysms were found in all passable explanted PHBV/PCL/GFmix$^{Hep/Ilo}$ grafts (figure 1).

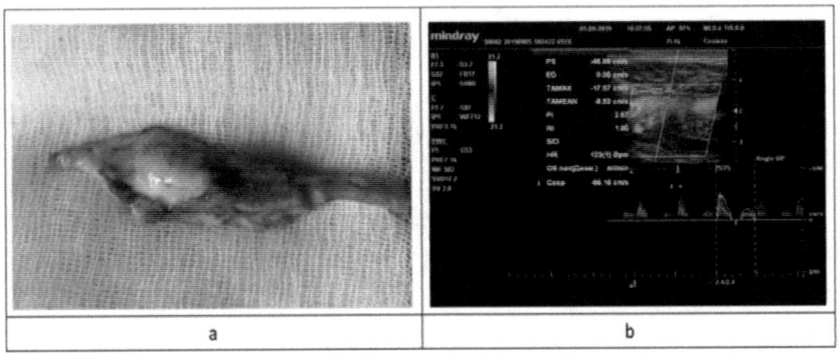

Figure 1. Image (a) and patency (b) of a PHBV/PCL/GFmix$^{Hep/Ilo}$ graft after 1.5 years of implantation.

A histological study of explanted samples showed that the biodegradable sheath of PHBV/PCL/GFmix$^{Hep/Ilo}$ thrombosed grafts was almost completely resorbed (figure 2). Formation of a newly formed three-layer vessel similar in structure to the native carotid artery of a sheep was revealed in the place of the biodegradable tubular sheaths. A layer of endothelial cells was visualized on the inner side of the neointima of the newly formed vessel. Media is represented by cells similar in morphology to smooth muscle cells. The smooth muscle cells were followed by a layer of connective tissue and adventitia, which contained a large number of newly formed vessels, lymphatic follicles and a small amount of perivascular adipose tissue. In addition, the adventitia structure contained a small number of fibroblast-like cells and foreign body cells. However, unlike the native vessel, the newly formed vascular tissue had no elastic fibers and a clear elongation of the cytoplasm of smooth muscle cells, which probably caused aneurysmal stretching of the newly formed vascular tissue due to pulsating blood flow. Additionally, a small focus of calcium was found under the neointima in the wall of one of the four passable grafts.

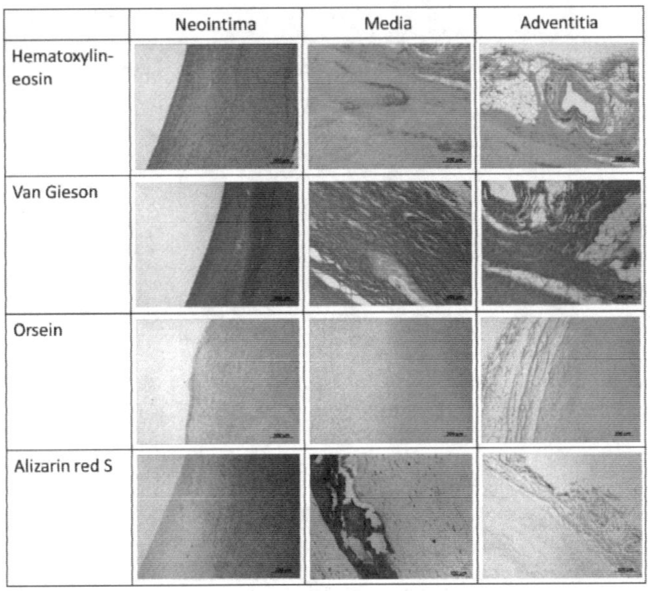

Figure 2. Results of histological examination of PHBV/PCL/GFmix$^{Hep/Ilo}$ grafts after 1.5 years of implantation, 100x magnification.

After 6 months of implantation, obturating recanalized thrombus and foci of crystalline calcium were detected in the lumen of Gore-Tex prostheses both in the wall of the prosthesis and in adjacent connective tissue capsule (figure 3).

IOP Conf. Series: Materials Science and Engineering 1093 (2021) 012017 doi:10.1088/1757-899X/1093/1/012017

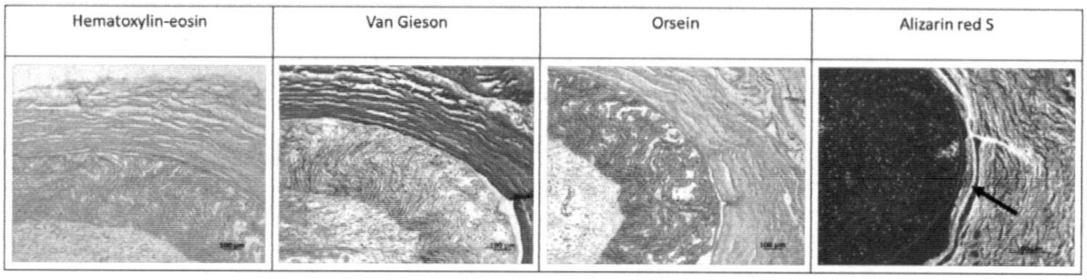

Figure 3. Results of histological examination of Gore-Tex grafts explanted 6 months following surgery. The arrow indicates the place with crystalline calcium in the capsule of the prosthesis, 100 x magnifications.

3.2. Results of the immunofluorescence study

The presence of endothelium containing signs of endothelial-to-mesenchymal transition (simultaneous expression of CD31 and α-actin, vWF) was noted in passable PHBV/PCL/GFmix$^{Hep/Ilo}$ grafts. The basal membrane contained type I and type IV collagen. Type III collagen is also found in the graft wall. The lumen of the Gore-Tex vessels was filled with a blood clot. The formation of newly formed tissue in the prostheses of this group was absent (figure 4).

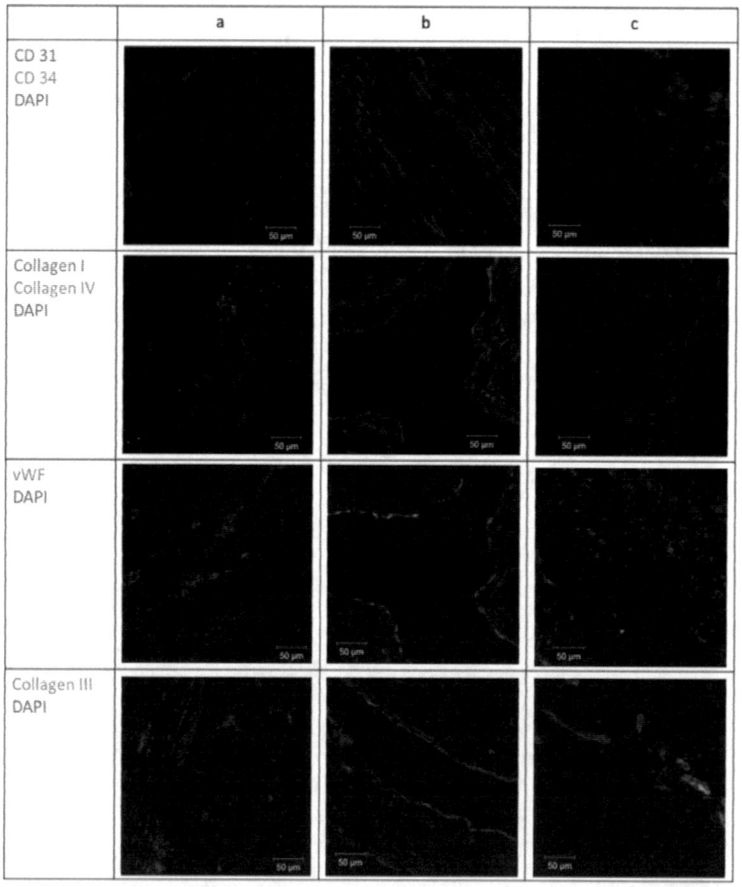

Figure 4. Comparison of the results of an immunofluorescence study of PHBV/PCL/GFmix$^{Hep/Ilo}$ prostheses and Gore-Tex prostheses. a - Gore-Tex; b – passable PHBV/PCL/GFmix$^{Hep/Ilo}$ prostheses; c – thrombosed PHBV/PCL/GFmix$^{Hep/Ilo}$ prostheses, 200x magnification.

4. Conclusion

The results of the study demonstrated high biocompatibility of PHBV/PCL/GFmix$^{Hep/Ilo}$ vascular prostheses compared to Gore-Tex prostheses during long-term implantation. The PHBV/PCL/GFmix$^{Hep/Ilo}$ biodegradable scaffolds tested in this study have shown the ability to assist with the formation of vascular tissue, but they require sheath reinforcement in order to prevent thrombosis. It is also necessary to further increase non-thrombogenous properties of the inner surface of the modified prostheses.

Acknowledgements
The work was performed within the framework of the program world-class Research and Education center "Kuzbass"

References

[1] Catto V, Fare S, Freddi G *et al* 2014 Vascular tissue engineering: recent advances in small diameter blood vessel regeneration *ISRN* **2014** pp1–27
[2] Taggart DP *et al* 2013 Current status of arterial grafts for coronary artery bypass grafting *Ann Cardiothorac Surg* **2** 4 pp 427–430
[3] Ren X, Feng Y, Guo J *et al* 2015 Surface modification and endothelialization of biomaterials as potential scaffolds for vascular tissue engineering applications *Chem. Soc. Rev* **44** 15 pp 5680–5742
[4] Antonova LV, Sevostyanova VV, Rezvova MA *et al* Technology of producing functionally active biodegradable small-diameter vascular prostheses with drug coating: patent # 2702239. Proprietor: Federalnoe gosudarstvennoe byudzhetnoe nauchnoe uchrezhdenie "Nauchno-issledovatelskij institut kompleksnykh problem serdechno-sosudistykh zabolevanij" (NII KPSSZ) (RU); № 2019119912; effective date for property rights: 25.06.2019; registration date: 07.10.2019, Bull. № 28

The technology for concrete production using an activated mixture of wood processing waste and sand

I N Kuznetsova

Yugra State University, Russia, Khanty-Mansiysk, 628012, Russia

E-mail: inkuznetsova-gsh@mail.ru

Abstract. The article presents a technology for producing concrete using the waste of wood processing plants as the concrete filler. The waste was activated by mechanical, hydromechanical and hydromechanochemical technologies in a centrifugal disk installation. The established optimal ratio of sand to the waste (activated mixture) is 70:30%. The accepted fraction of the activated mixture is 10 wt% (of the mass of cement). The tests confirmed the reduction in the average density of the cement stone, while preserving its strength; the thermophysical properties of the stone increase.

1. Introduction

The concrete based on the waste of wood processing plants is a structural and heat insulating material. Novel technologies and methods that increase the characteristics of binders due to admixture of nanofunctional additives in combination with nanotechnologies enable production of concretes with predefined properties intended for specific operation conditions.

The implemented waste has different moist content: dry (below 15%) and semi-dry (from 16 to 30%). In terms of their structure, shavings and wood dust (as an ultrafine filler) are classified as bulk materials. The surface-active properties of wood increase with its fineness. The average wood density largely ranges from 380 to 1100 kg/m^3 in a completely dry state. The bulk density of a dry hogged chips of pine is 675 kg/m^3, that of aspen is 554 kg/m^3, that of birch is 247.2 kg/m^3. The thermal conductivity along the tree wool is 0.2–0.3 W/(m·°C). Across the wool, it is considerably smaller: 0.11–0.17 W/(m·°C). For ready wooden-concrete product with an average density of 830 kg/m^3, the thermal conductivity coefficient is 0.32 W/(m·°C).

A specific feature of the technologies of building materials on the basis of dispersed system is constant presence of a solid phase in the system, regardless of the presence of liquids and gases. The solid phase is the carrier of strength — the basic property of construction materials; it takes part in the formation of all the three structures mentioned above. The solid phase particles are elementary "blocks" that form various structures which strength is predetermined, primarily, by their dispersity and granulometric composition. The smaller the particles, the fewer internal defects they contain. The presence of such particles in base dispersed systems is mandatory. The structure of the base dispersed systems (cement paste and mortars, plaster paste, molding compounds and mixtures, various suspensions, etc.) should distinguish two components: structural matrix formed by coarse particles and pore substance composed of fine particles and liquid phase inside the porous volume of the coarse matrix. Colloid-chemical properties of the pore substance and its content in the system have the decisive impact on the strength formation of different structures. [1]

The study was also aimed at optimizing the activated composition of the mixture of sand (as abrasive material) and wood processing waste and its mass content in the concrete to reveal the effect of the activated mixture (sand : wood processing waste) on the physicomechanical properties of concrete. The results were used to propose a technology of concrete production using the wood processing waste. [2, 5-7]

2. Experimental

The centrifugal disk installation was used to activate the waste by mechanical action (dry milling) allowed obtaining ultrafine materials; hydromechanic activation (wet milling) of the material in the installation enabled hydrodynamic, acoustic and turbulent cavitation treatment of the mixture; hydromechanochemical activation (wet milling and chemical reagent), chemical reagent: plasticizing additive "Megalit S-3 ML", powder density $\rho = 1.15$ kg / m^3.

The mechanical and hydromechanical activation yielded ultrafine materials with the size as small as 0.001 nm.

The weight distribution of the particle dimensions (figure 1) was determined on a MicroSizer–201 laser analyzer (VA INSTALT, Russia): 70% of sand and 30% of wood processing waste.

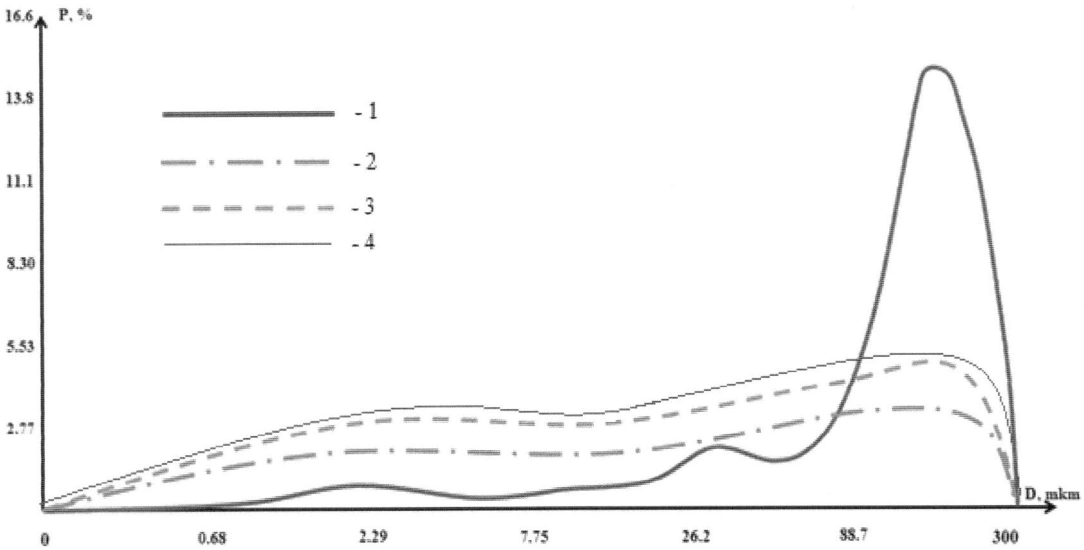

Figure 1. Weight particle distribution of the activated mixture:
1 - Designed values (no activation); 2 - Mechanical activation;
3 - Hydromechanical activation; 4- Hydromechanochemical activation.

The production of concrete with wood processing waste includes the following technological stages:

1. Optimization of the disk mill-activated mixture: 70% of sand with 30% of wood processing waste.
2. The filler (sand with ultrafine filler) and cement (TsEM II/A-Sh 42.5N) are dosed and mixed for 5–6 minutes.
3. The concrete is poured into molds with preliminary holding of the concrete mixture over 3–4 hours at a temperature of 20 °C.
4. Thermal-moist treatment in an autoclave at 12 atm. in the following mode: increase of the temperature up to 80 °C for 4 hours, isothermal heatup at 80 °C for 6 hours; temperature reduction down to 20 °C for 4 hours.

Based on the results of activation of the concrete mixture, the optimal mass ratio of sand to the wood processing waste was accepted as 30:70%, based on the acceptable variation of the strength of the control specimens (4–5%), while their thermophysical properties improved.

The physicomechanical properties of the cement stone specimens made from cement-sand-waste mixture are presented in Table 1.

Table 1. Physicomechanical properties of cement stone specimens.

Physicomechanical indicators of activated mixture (70% of sand/30% of wood processing waste)	Specific surface S_{sp} [cm²/g]	Powder density ρ [g/cm³]	Average particle size D [μm]	Bending strength R_{bend} [MPa]	Compression strength R_{compr} [MPa]	Thermal conductivity coefficient λ [W/(m·°C)]
Designed values (no activation)	3450	218	47,0	16.7	42,0	0.46
Mechanical activation	6980	318	5.3	17.8	45.5	0.42
Hydromechanical activation	7445	350	4.2	18.5	46.2	0.40
Hydromechanochemical activation	8017	355	4.0	18.7	46.5	0.39

The strength indicators of the cement stone specimens from the activated mixture (sand:waste = 70%/30%) with 10 wt% of cement are presented in Table 2.

Table 2. Strength of specimens.

Physicomechanical indicators of activated mixture (70% of sand/30% of wood processing waste) with 10 wt% of cement	Bending strength R_{bend} [MPa]	Compression strength R_{compr} [MPa]
Mechanical activation	17.4	45.8
Hydromechanical activation	18.2	46.5
Hydromechanochemical activation	18.4	46.7

The structure of the specimens was determined using a REM-100U scanning electron microscope (Russia). Figure 2 illustrate the microstructure of pore space and pore walls of the specimens, respectively.

Figure 2. Microstructure of pore space and pore walls of foam concrete.

Evidently from the figures, needle-like ettringite crystals penetrate calcium hydrosilicate particles, which contributes to the strengthening of the microstructure and augments the characteristics of the foam concrete specimens as a whole.

The mineral composition of the specimens was analyzed on a D8 ADVANCE XRD analyzer (Bruker, USA). The analysis has shown that the cement stone specimens of cement TsEM II/A-Sh 42.5N contain the following minerals: calcium carbonate $CaCO_3$, calcium hydroxide $Ca(OH)_2$, calcium hydrosilicate Ca_2SiO_4 and calcium alumosilicate $Ca_3Al_2Si_3O_{12}$. The thermal conductivity of the specimens based on the same cement without addition of wood processing waste was 0.46 W/(m·°C). [3, 4, 6]

3. Conclusion

The article presents a technology for producing concrete using the waste of wood processing plants as the concrete filler. The waste was activated by mechanical, hydromechanical and hydromechanochemical technologies in a centrifugal disk installation.

The introduction of different fraction of shavings and wood dust decreased the thermal conductivity coefficient of the cement stone specimens (Table 2). With larger fraction of shavings and wood dust in the specimens, the strength decreases by 5–7% for each 4–5% of filler; the thermal conductivity coefficient reduces by 7–9%.

The established optimal ratio of sand to the waste (activated mixture) is 70:30 %. The accepted fraction of the activated mixture is 10 wt% (of the mass of cement). The decrease of sand-to-waste ratio is unreasonable because this reduces the strength of the cement stone by 2.5% and more.

The application of ultradispersed wood processing waste in concrete as an activated mixture (70% of sand + 30% of waste) allows reducing the average density by 10% without deterioration of strength and increasing the thermophysical characteristics of the cement stone.

The tests confirmed the reduction in the average density of the cement stone, while preserving its strength. The thermophysical properties of the stone increase.

The use of the suggested innovative construction material from easily available waste of wood processing plants for civil and industrial building may facilitate the attraction of considerable amount of investment, which will be economically and socially beneficial for the country.

References

[1] Lotov V A 2007 Nanodisperse systems in construction materials technology *Proceedings of Tomsk Polytechnic University* 311 **3** 84-88

[2] Verba C, Montross S, Spaulding R, Dalton L, Crandall D, Moore J, Glosser D, Huerta N, and Kutchko B 2017 Foamed Cement Interactions with CO_2 *United States* Web. doi:10.2172/1373376

[3] Luo X, Dong J, Zhang L, Du J, Wang H and Gao W 2017 Preparation of silica micro spheres via a semibatch sol–gel method *J. Sol-Gel Sci. Technol.* **81** 669-77

[4] Kearsley E P, Wainwright P J 2002 The Effect of Porosity on the Strength of Foamed Concrete *J. Cem. Concr. Res.* **32** 233–239

[5] Haljasmaa I, Kutchko B, Spaulding R, Crandall D, Moore J, Gill M, Benge G *et al* 2019 Foamed Cement: Correlation of Foam Quality with Mechanical and Physical Properties *United States* Web. doi:10.18141/1506981

[6] I N Kuznetsova *et al* 2019 Technology of foam concrete production with ultradispersed quartz waste *IOP Conf. Ser.: Mater. Sci. Eng.* 597 012031. Web. doi.org/10.1088/1757-899X/597/1/012031

[7] M A Darulis et al. 2019 Structure and properties of the cement stone modified by ultradispersed quartz waste *IOP Conf. Ser.: Mater. Sci. Eng.* 597 012032. Web. doi.org/10.1088/1757-899X/597/1/012032

Damage accumulation after temperature aging in structural GFRP in interlayer shear tests

D S Lobanov and E M Zubova

Centre of Experimental Mechanics, Perm National Research Polytechnic University, 29 Komsomolsky Ave., Perm 614000, Russia

E-mail: cem.zubova@mail.ru

Abstract. A series of quasi-static tests of the structural fiberglass reinforced polymers samples in interlayer shear by the short-beam method were carried out in this work. Mechanical tests were carried out after preliminary temperature aging with different temperature regimes and exposure duration. Diagrams of dependence of the residual interlayer strength degradation on the mode of preliminary aging are constructed. An investigation of the microstructure of GFRP specimens before and after quasi-static tests in interlayer shear was carried out. During the experimental studies the control of the damage accumulation processes and destruction was carried out using the acoustic emission method (AE). Analysis of the AE signal parameters made it possible to assess the prevalence of different destruction types of composites depending on the mode of preliminary aging.

1. Introduction

The conditions of the exploitative environment have a great influence on the destruction processes of composite materials. It is important to evaluate the duration and level of thermal effects in the design of structural elements as well as the features of the material of the real construction. In this regard the problem of studying the features of mechanical behavior, deformation pattern and destruction of structural composites during preliminary temperature aging, thermal moisture aging and moisture saturation in operating (polluting) environments becomes more actuality. In addition, to obtain characteristic dependences of changes in mechanical properties on the corresponding mode of preliminary action is important. [1-6].

Thus, the experimental study of the effect of elevated temperatures on the processes of damage accumulation and destruction of structural FGRP used for the manufacture of critical elements and structures is a relevant intervention of research in the solid mechanics.

2. Experimental procedure

In this work, an experimental study of samples of structural fiberglass/epoxy "STEF" after temperature aging at 120°, 200°C for 5 and 15 days is carried out. Mechanical tests for interlayer shear under static conditions were carried out using the short beam method on the basis of the shared research facilities “Center of Experimental Mechanics” PNRPU. Mechanical tests were carried out with the recommendations of ASTM D2344 on an Instron 5965 electromechanical testing system. The loading rate was 1 mm / min.

In order to study the influence of preliminary temperature aging on the mechanisms of destruction and the stages of damage accumulation in samples of various structural composite materials, the method of acoustic emission (AE) was used in this work [7-11]. The energy of AE was chosen as an informative parameter of the signals (E, V^2s). In addition, the values of the frequencies of maximum of spectrum (F, kHz), obtained using the fast Fourier transform, were estimated. AE signals were recorded using the Vallen AMSY-6 system (Germany). A broadband piezoelectric sensor with a frequency range of 300-800 kHz and a preamplifier with a gain of 34 dB were used. Cyanoacrylate glue was used to attach the sensors to the sample surface. A photo of a sample with an AE sensor is shown in figure 1.

The processes of damage accumulation and mechanisms of destruction of GFRP during mechanical loading are studied using the AE method. The expediency of using the frequency characteristics of the AE signal are noted. The most commonly used parameter is the values of the frequencies of maximum of spectrum [9-10, 12-15]. When analyzing these works, it was noted that the authors determined the frequency ranges of the spectrum maximum to describe the main mechanisms of damage accumulation in fiberglass/epoxy (matrix cracking, adhesion between the fiber and the matrix, fiber destruction). It is noted that lower values of the frequency ranges correspond to matrix cracking, high values characterize fiber destruction, and intermediate ones are associated with disruption of adhesion between the fiber and the matrix and delamination.

Figure 1. A photo of a sample with an AE sensor.

3. Result and discussions

3.1. Mechanical test results

The results of mechanical tests of structural fiberglass/epoxy samples in interlayer shear using the short beam method are shown in figure 2. An increasing dependence of the change in the residual ultimate strength during interlayer shear on the selected mode of preliminary temperature aging to the group of samples without thermal exposure (nominal) is noted.

Studies of the microstructure after temperature aging of the fiberglass/epoxy "STEF" samples before and after interlayer shear tests showed that at 120 °C - 5 days/15 days the matrix color changes to more yellow (5 days) and brown (15 days), there are no visible structural defects before testing. At 200 °C - 5 days/15 days the matrix color change to dark gray/black, defects in the form of surface cracks appear along the fiber-matrix boundaries.

IOP Conf. Series: Materials Science and Engineering 1093 (2021) 012019 doi:10.1088/1757-899X/1093/1/012019

After testing the samples of fiberglass/epoxy “STEF”, it can be noted that for a group of samples 120 °C - 5 days a macrocrack is formed due to shear of the layers. At 120 °C - 15 days, the destruction of the binder leads to the formation of many interlayer cracks, the development of which leads to the destruction of the sample. At 200 °C - 5 and 15 days, destruction and local brittleness of the binder and the formation of micro cracks occur, followed by shear failure of the samples. However, it should be noted that temperature aging lead to an increase in stiffness and an increase in strength values at interlayer shear by 15-30%, depending on the mode (figure 2).

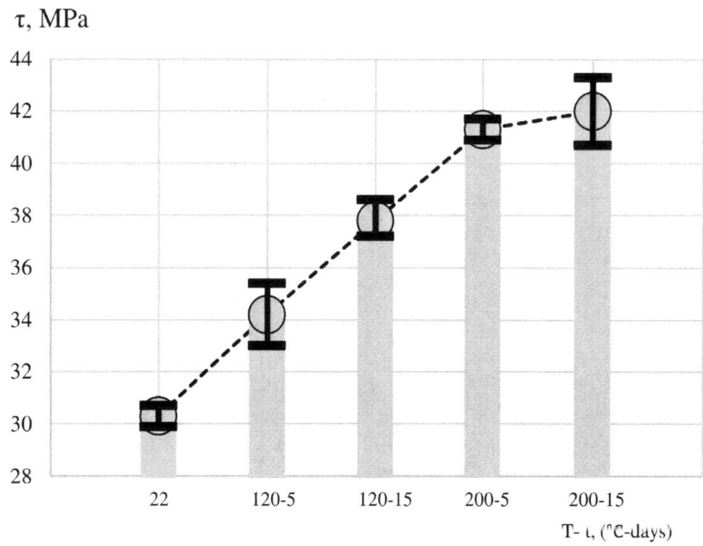

Figure 2. Dependences of the change in the residual tensile strength at interlayer shear on the selected pre-exposure mode for fiberglass/epoxy “STEF” specimens.

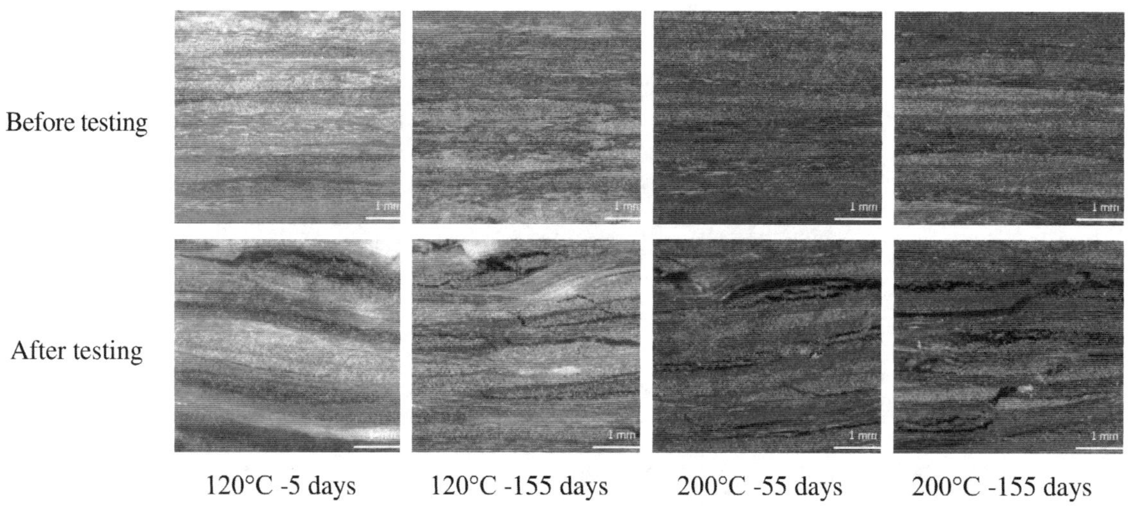

Figure 3. Structure of fiberglass/epoxy specimens after temperature aging under different mode before and after tests.

3.2. Acoustic Emission Signal Processing

To analyze the effect of preliminary temperature aging on the damage accumulation processes in samples of structural fiberglass, graphs of the distribution of the values of the frequencies of maximum of spectrum for all groups of samples under static load were plotted (figure 4).

For the "STEF" samples, it was noted that the values of the frequencies of maximum of spectrum can be divided into three ranges (25-80 kHz, 270-330 kHz and > 650 kHz). For samples tested after preliminary aging at a temperature of 200° C and holding for 5 and 15 days, the largest number of recorded signals was recorded in the low frequency range (25-80 kHz). For the samples tested after preliminary aging at a temperature of 120 ° C and holding for 5 and 15 days, the largest number of registered signals was recorded in the middle frequency range (≈ 300 kHz).

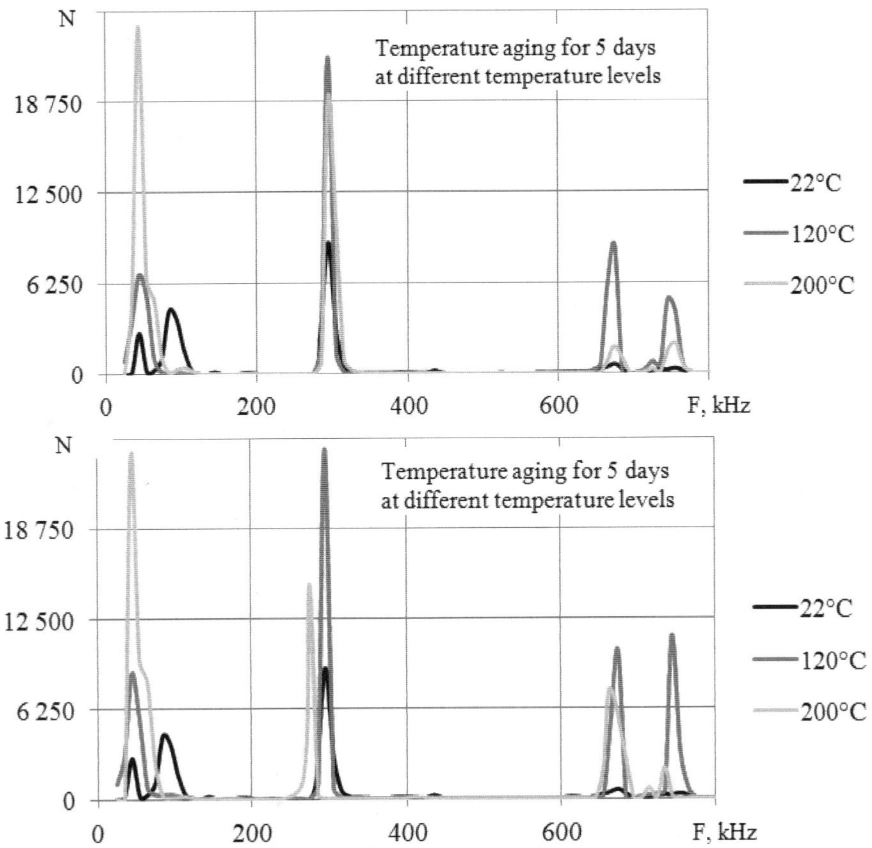

Figure 4. Graphs of the distribution of the values of the frequencies of maximum of spectrum for all groups of samples under static load.

4. Conclusions

New experimental data that illustrate the effect of preliminary temperature aging at elevated temperatures on the staging of damage accumulation in fiberglass samples during static tests in interlayer shear by the short beam method have been obtained. Non-monotonic dependences of the change in the residual ultimate strength during interlayer shear on the selected mode of preliminary temperature aging are noted. The most dangerous modes of temperature aging are determined, which lead to a change in the mechanisms of destruction and accumulation of damage and limit the implementation of the deformation and strength properties of fiberglass/epoxy "STEF". Was noted that the values of the frequencies of maximum of spectrum for all samples can be divided into three ranges and related to the main mechanisms of destruction in composites.

Acknowledgments

The reported study was funded by RFBR, project numbers 19-31-90148, 19-41-590005 r_a. The Experimental studies in the framework of the description of the mechanical behavior of structurally inhomogeneous materials were conducted within the State Assignment of the Ministry of Science and Higher Education of the Russian Federation (№ FSNM-2020-0027).

References

[1] Alam P, Robert C and Ó Brádaigh C M 2018 Tidal turbine blade composites - A review on the effects of hygrothermal aging on the properties of CFRP *Composites Part B* **149** pp 248-259
[2] Lobanov D S, Vildeman V E, Babin A D and Grinev M A 2015 Experimental research into the effect of external actions and polluting environments on the serviceability of fiber-reinforced polymer composite materials *Mech. Compos. Mater.* **51** pp 69-79
[3] Mansouri L, Djebbar A, Khatir S and Abdel Wahab M 2019 Effect of hygrothermal aging in distilled and saline water on the mechanical behaviour of mixed short fibre/woven composites Compos. Struct. **207** pp 816-825
[4] Nicholas J, Mohamed M, Dhaliwal G S, Anandan S and Chandrashekhara K 2016 Effects of accelerated environmental aging on glass fiber reinforced thermoset polyurethane composites *Composites Part B* **94** pp 370-378
[5] Lobanov D S, Slovikov S V 2018 Mechanical behavior of a unidirectional basalt-fiber-reinforced plastic under thermomechanical loadings *Mech. Compos. Mater.* **54** pp 351-358
[6] Slovikov S V and Lobanov D S 2020 Mechanical Properties of a basalt-fiber-reinforced plastic rod used in composite high-voltage wires in torsion and three-point bending *Mech. Compos. Mater.* **56** pp 353-360
[7] Lobanov D S and Zubova E M 2020 Temperature aging effects on mechanical behavior of structural GFRP on interlaminar shear tests *IOP Conf. Series: Materials Science and Engineering* **747** 012119
[8] Lobanov D S and Zubova E M 2019 Research of temperature aging effects on mechanical behaviour and properties of composite material by tensile tests with used system of registration acoustic emission signal *Procedia Structural Integrity* **18** pp 347-352
[9] Arumugam V, Saravanakumar K and Santulli C.2018 Damage characterization of stiffened glass-epoxy laminates under tensile loading with acoustic emission monitoring *Composites Par B* **147** pp 22-32
[10] Barile C, Casavola C, Pappalettera G and Vimalathithan K 2019 Damage characterization in composite materials using acoustic emission signal-based and parameter-based data *Composites Part B* **178** 107469
[11] Beheshtizadeh N, Mostafapour A and Davoodi S 2019 Three point bending test of glass/epoxy composite health monitoring by acoustic emission *Alexandria Engineering Journal* **58** pp 567-578
[12] Oskouei A R, Heidary H, Ahmadi M and Farajpur M 2012 Unsupervised acoustic emission data clustering for the analysis of damage mechanisms in glass/polyester composites *Materials & Design* **37** pp 416-422
[13] Tabrizi I E, Kefal A, Zanjani J S M, Akalin C and Yildiz M 2019 Experimental and numerical investigation on fracture behavior of glass/carbon fiber hybrid composites using acoustic emission method and refined zigzag theory *Compos. Struct* **223** 110971
[14] Saravanakumar K and Arumugam V 2018 Effect of milled glass fibers on quasi-static indentation and tensile behavior of tapered laminates under acoustic emission monitoring *Engineering Fracture Mechanics* **201** pp 36-46
[15] Beheshtizadeh N, Mostafapour A and Davoodi S 2019 Three point bending test of glass/epoxy composite health monitoring by acoustic emission *Alexandria Engineering Journal* **58** pp 567-578

Microstructure, electromechanical hysteresis and polarization characteristics of solid solutions of the system (Na, K, $Cd_{0.5}$) NbO_3

M O Moysa[1], K P Andryushin[1], L A Shilkina[1], A V Nagaenko[2], I N Andryushina[1] and L A Reznichenko[1]

[1]Southern Federal University, Scientific Research Institute of Physics, Stachki av. 194, 344090, Rostov-on-Don, Russia
[2]Southern Federal University, Institute of High Technology and Piezo Technic, st. Milchakova 10, 344090, Rostov-on-Don, Russia

E-mail: maksim.moysa@mail.ru

Abstract. The phase diagram of the system (1-x-y) $NaNbO_3$ – $x KNbO_3$ – $y CdNb_2O_6$ (y=0.25, x=0.05÷0.20) is constructed. The agreement between the microstructure and the phase diagram of the system is demonstrated. A decrease in the polarization characteristics up to the phase transition temperature is detected. The formation of extrema $d_{33}(E)$ in the region of the inflection point on the ascending branch of the half-period S_{33} is revealed. A correlation has been established between the internal structure (crystalline, domain, grain, defect), fundamental electromechanical and polarization properties for solid solutions of the studied system.

1. Introduction

The most widely used piezoelectric materials are solid solutions based on system $PbTiO_3$-$PbZrO_3$ (PZT). This is due to their excellent piezoelectric properties [1-3]. The development of the legislative framework in the field of environmental protection is forcing the search for lead-free materials that can replace PZT ceramics. This will allow to reduce anthropogenic pollution. Among the latter, a special place is occupied by materials based on solid solutions (SS) of sodium niobate (SN) [1, 4-6], which have a unique combination of properties: low specific gravity ($\rho \leq 4.5$ g/cm^2), high velocity of sound ($v_R \sim 6$ km/s), a wide range of values of relative dielectric constants ($50 \leq \varepsilon_{ij}/\varepsilon_0 \leq 2000$) and mechanical quality factor ($50 \leq Q_M \leq 2000$) at sufficiently high piezoelectric parameters, the achievement of which in PZT materials is fundamentally impossible.

At the same time, niobate materials have not yet found wide application in equipment, while the number of known SS systems containing PZT has reached several hundred. Among the latter, multicomponent systems have become the basis of materials for various purposes. The range of studied systems based on SN is limited no more than 50 compositions, which not sufficiently investigated and only a few binary systems have found real practical application. This is due to the impossibility of obtaining solid solutions based on SN by traditional methods, which, is associated with the complexity of their crystal structure, the presence of a large number of phase transitions, and the strong dependence of properties on the conditions of structure formation.

This work continues our earlier studies of the system (1-*x*-*y*) $NaNbO_3$ – *x*$KNbO_3$ – *y*$Cd_{0.5}NbO_3$ [7] and is devoted to the establishment of regularities in the formation of the phase picture, grain landscape (microstructure), electromechanical and polarization properties of the SS system in a wide range of external influences (temperatures, constant and alternating electric fields).

2. Experiments

The objects of the study were SS of the composition (1-*x*-*y*) $NaNbO_3$ – *x*$KNbO_3$ – *y*$CdNbO_3$ system with y=0.25, *x*=0.05÷0.20, Δ*x*=0.05. The samples were prepared by two-stage solid-phase synthesis and sintering using conventional ceramic technology. ($T_{synth.1}$=1220 K, *τ*=5 h., $T_{synth.2}$= 1240 K, *τ*=10 h.; $T_{sint.}$=(1190÷1200) K depending on the composition). $NaHCO_3$ ("chemically pure"), $KHCO_3$ ("pure for analysis") and Nb_2O_5, CdO ("pure") were used as the feedstock.

Sintered ceramic blanks were subjected to mechanical processing (cutting along the plane, grinding along flat surfaces and ends) in order to obtain measuring samples of Ø10mm x 1mm. Before metallization, the samples were calcined at a temperature $T_{calc.}$=770 K for 0.5 h. to remove organic matter residues and degrease surfaces in order to increase the adhesion of the metal coating to ceramics. The electrodes were applied by double firing a silver-containing paste at a temperature of 1070 K for 0.5 h.

X-ray diffraction studies were performed at room temperature using a DRON-3 diffractometer (filtered Co_{K_α}-radiation, Bragg - Brentano focusing scheme). Bulk and ground ceramic objects were studied, which made it possible to exclude the influence of surface effects, voltage and textures arising in the process of obtaining ceramics. The calculation of structural parameters was performed according to standard methods [8].

The obtained samples were polarized using the "hot" polarization method. In this case, the samples were loaded into the chamber with PES-5 polyethylene siloxane liquid at ~ 300 K for 0.5 h., then the temperature was smoothly risen to 423 K, while the field was simultaneously increased from 0 to (3.0÷4.0) kV/mm. Under these conditions, the samples were kept for (20÷25) minutes and then cooled under the field to ~300 K (room temperature).

A scanning electron microscope JSM – 6390L (JEOL) (Japan) with a system of microanalyzers of the Oxford Instruments company (Great Britain) was used to study the microstructure of sample chips; microscope resolution was up to 1.2 nm at an accelerating voltage of 30 kV (an image in secondary electrons), accelerating voltage limits were from 0.5 to 30 kV, magnification range: from x5 to x300000, beam current was up to 200 nA.

The P-E loops were studied using the Sawyer - Tower scheme, at f=50 Hz and the temperature range of (290÷431) K. The values Ps – the saturation polarization, Pr – the residual polarization, Ec – the coercive field were estimated. The narrow temperature range was conditioned by the use of polyethylene siloxane oil with a boiling point of ≈480 K as an insulator in the measuring cell. To eliminate the effect of hydrolization, the samples were preliminarily annealed at T=670 K.

The curves of unipolar deformation (S_{33}) induced by a constant electric field with intensity E and (S_{33}-E) were recorded on polarized specimens using a specially fabricated test stand at the Research Institute of Physics, Southern Federal University, based on a MIKRON-2 instrument for verifying end gauges, an Agilent 34420 A nanovoltmeter / microohmmeter. The curves of $S_{33}(E)$ were approximated using the least squares approach. Minimum standard deviation r was our criterion of the quality of approximation. Inverse piezomodulus d_{33} was calculated using the equation $d_{33}=S_{33}/E$.

3. Results and discussion

Figure 1 shows the phase diagram (PD) of the system with *y*=0.25, which indicates the formation of regions with coexisting phases with different symmetry.

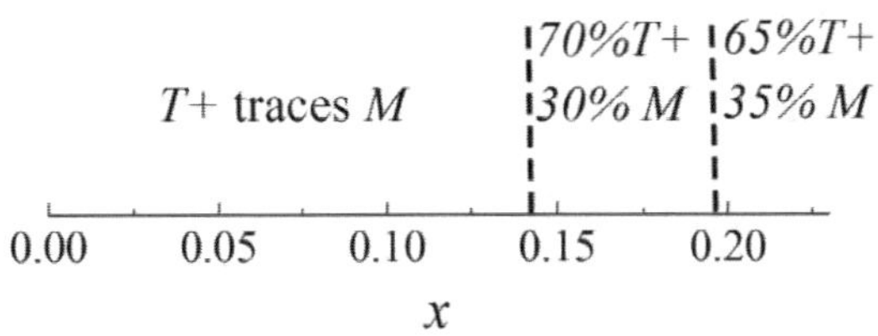

Figure 1. Phase diagram of the (1-x-y) $NaNbO_3$ – $x$$KNbO_3$– $y$$Cd_{0.5}NbO_3$ y=0.25, x=0.05÷0.20.

It is seen that near $NaNbO_3$ ($x \leq 0.15$) tetragonal (T) structures with traces of monoclinic (M) - structures crystallize. An increase in the content of the $KNbO_3$ ($0.15 \leq x \leq 0.25$) leads to the formation of a heterogeneous region with coexisting T- and M- phases. Recrystallization processes occurring against the background of such phase transformations lead to the formation of a kind of grain landscape (Figure 2), which clearly responds to all changes in the crystal structure of the SS. From figure 2 it is clearly seen that the microstructure of ceramics with a very low content of $KNbO_3$ (x = 0.05 ÷ 0.10) is close to homogeneous (region I, RI). At $0.15 \leq x \leq 0.20$, the microstructure is refined, becomes inhomogeneous, and regions with coarse and fine grains appear, the number of which (small) increases with approaching $KNbO_3$ (region II, RII).

Figure 2. Fragments of the ceramic microstructure of SS of the (1-x-y) $NaNbO_3$ – $x$$KNbO_3$– $y$$Cd_{0.5}NbO_3$ y=0.25, x=0.10÷0.20.

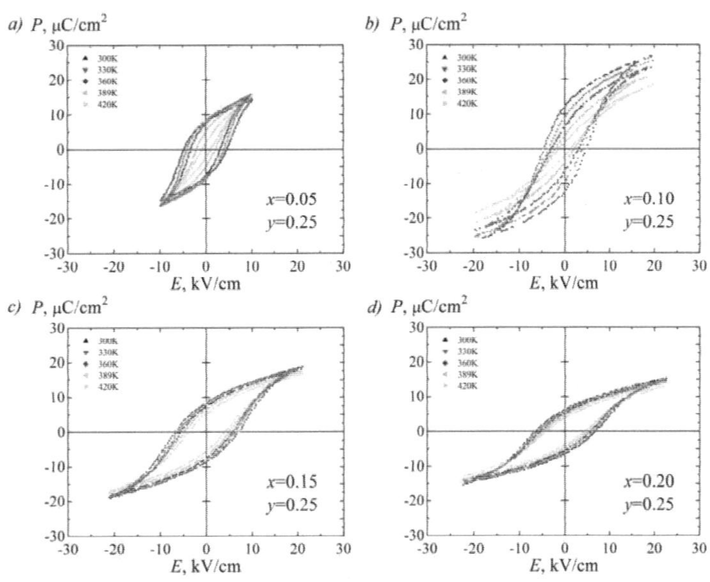

Figure 3. *P-E* loops of SS of the composition (1-x-y) $NaNbO_3$ – $x$$KNbO_3$– $y$$Cd_{0.5}NbO_3$ (y=0.25, $0.05 \leq x \leq 0.20$) at T=300÷420K, ΔT=30K.

Figure 3 shows the *P-E* loops of the studied SS at T=300÷420K, ΔT=30K. It was revealed that, in a SS with x = 0.05, the formation of saturated *P-E* loops occurs at a voltage of U = 700 V over the entire investigated temperature range (Figure 3*a*). Exceeding U = 700V causes the loop to expand and, consequently, increase energy losses during the polarization reversal cycle. In SS with $0.10 \le x \le 0.20$, the permissible voltage at which typical ferroelectric $P - E$ loops are formed is $U = (1600 \div 1800)$ V (Figure 3 *b-d*). An increase in temperature leads to a gradual contraction of the loop along the ordinate axis.

Based on the study of the *P-E* loops, the dependences of (P_s, P_r, E_c) (T) (Figure 4) have been constructed and approximated by a polynomial of the second degree. It was found that at x =0.05, the inflection point P_s, P_r of characteristics is formed in the temperature range ~ (360 ÷ 370) K. Up to the indicated temperatures (P_s, P_r)(T), the dependences either increase slightly (Figure 4) or do not change. Parameter E_c decreases over the entire research temperature range. The observed anomaly may indicate structural instabilities present in these objects. At $0.15 \le x \le 0.20$, the dependences (P_s, P_r, E_c) (T) tend to linearly decrease with increasing temperature.

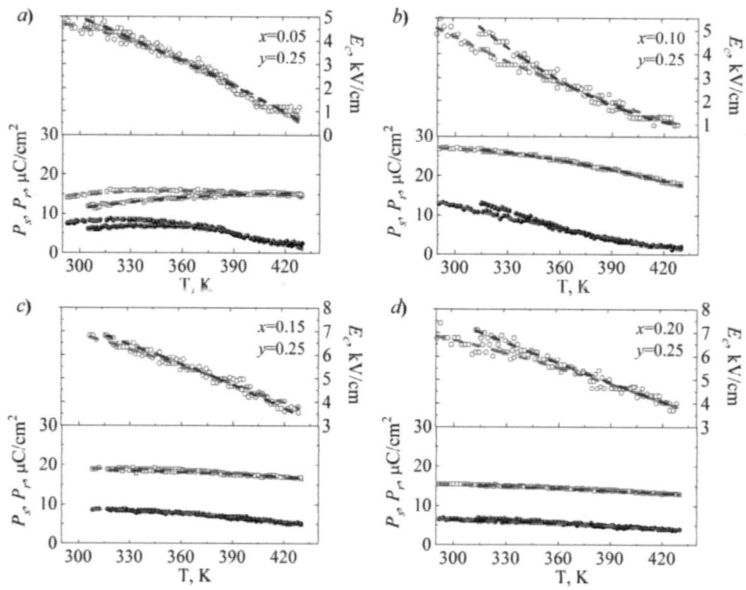

Figure 4. Dependences of (P_s, P_r, E_c)(T) of SS of the composition (1-x-y) $NaNbO_3 - xKNbO_3 - yCd_{0.5}NbO_3$ (y=0.25, $0.05 \le x \le 0.20$).

Figure 5 illustrates the evolution of S_{33}-*E* loops depending on the concentration of $KNbO_3$. It was revealed that in the investigated SS the dependence $S_{33}(E)$ is characterized by a shape close to the dielectric hysteresis loop. It is also shown that at x> 0.10, a gradual narrowing of the loops occurs over the entire range of constant electric field strengths. The resulting loop shape of the investigated SS at low concentrations x may be due to the softness of the structure, and their narrowing with enrichment in potassium niobate is probably a consequence of an increase in the rigidity of the latter and an increase in the heterogeneity of these objects.

From the ascending loop of electromechanical hysteresis, curves $d_{33}(E)$ were obtained at $0.05 \le x \le 0.20$ (Figure 6). It was revealed that in SS at $0.05 \le x \le 0.10$, the $d_{33}(E)$ curves are characterized by a sharp increase in d_{33} at low values of the constant electric field strength, followed by the formation of extrema in the inflection point area on the ascending branch of the half-period S_{33} [9]. For x> 0.10, the $d_{33}(E)$ curves in the region of low fields do not form extrema and increase smoothly. It should be noted that the curves $d_{33}(E)$ characterizing SS with $0.05 \leq x \leq 0.10$ and $x = 0.20$ reach saturation, i.e. have a plateau-like region. The $d_{33}(E)$ ($x = 0.15$) does not reach saturation in the entire research range of field strengths.

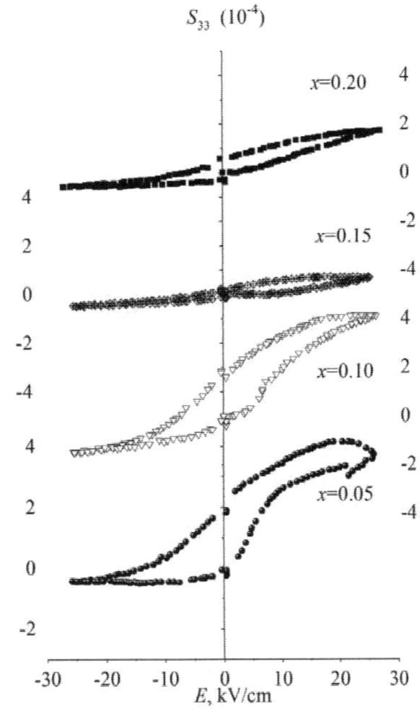

Figure 5. S_{33}-E loops of SS of the composition (1-x-y) $NaNbO_3$ – $x$$KNbO_3$– $y$$Cd_{0.5}NbO_3$ (y=0.25, $0.05 \leq x \leq 0.20$) at T=300K.

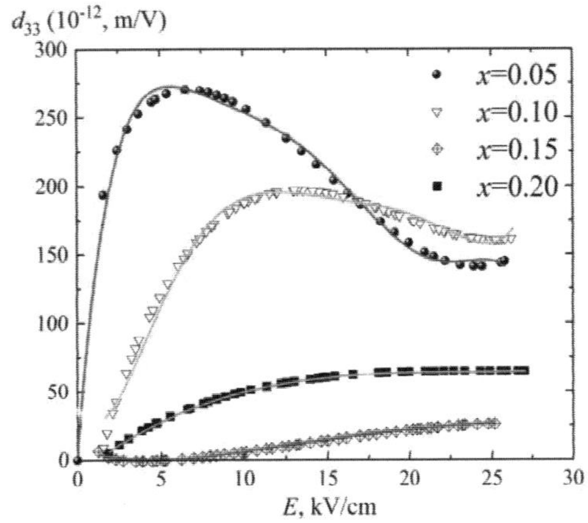

Figure 6. Curves d_{33} (E) of the SS composition (1-x-y) $NaNbO_3$ – $x$$KNbO_3$– $y$$Cd_{0.5}NbO_3$ (y=0.25, $0.05 \leq x \leq 0.20$).

4. Conclusions

Solid solutions of the three-component system (1-x-y) $NaNbO_3$ – $x$$KNbO_3$ – $y$$Cd_{0.5}Nb_2O_6$ (y=0.25, x=0.05÷0.20) were obtained. A diagram of the states of the system is constructed. The formation of regions with coexisting phases with different symmetry is detected. The matching of the microstructure and phase diagram of the system is demonstrated. The regularities of the formation of electromechanical and polarization properties of the SS system in a wide range of external influences (temperature, constant and alternating electric fields) have been established. It was found that at $0.05 \leq x \leq 0.10$, the inflection point P_s -, P_r- characteristics are formed in the temperature range ~ (360 ÷ 370) K. It was revealed that in SS at $0.05 \leq x \leq 0.10$, the $d_{33}(E)$ curves are characterized by a sharp increase in d_{33} at low values of the constant electric field strength, followed by the formation of extrema in the inflection point area on the ascending branch of the half-period S_{33}. A scientific explanation of the observed phenomena is given.

The obtained results should be considered in the development piezoelectric devices for various applications.

Acknowledgments
Research was financially supported by the Ministry of Science and Higher Education of the Russian Federation (State assignment in the field of scientific activity, Southern Federal University, 2020).

References

[1] Jaffe B, Cook W, Jaffe H *et al* 1971 Piezoelectric Ceramics *Academic Press*
[2] Reznichenko L A, Verbenko I A, Shilkina L A *et al* 2018 Binary, Ternary and Four-Component Systems Based on Sodium Niobate: Phase Diagrams of States, the Role of the Number of Components and Defectiveness in the Formation of the Properties *Advanced Materials. PHENMA* **207** pp 3-23
[3] Cekas E, Janusas G, Guobiene A, Palevicius A *et al* 2018 Design of Controllable Novel Piezoelectric Components for Microfluidic Applications *Sensors* **18** 11 4044
[4] Safari A, Hejazi M *et al* 2012 Lead-Free KNN-Based Piezoelectric Materials In: Priya S., Nahm S. (eds) Lead-Free Piezoelectrics *Springer New-York.* Chapter 5 pp 139-175
[5] Egerton L, Dillon D M *et al* 1959 Piezoelectric and Dielectric Properties of Ceramics in the System Potassium—Sodium Niobate *Journal of the American Ceramic Society* **42** pp. 438-442
[6] Smiga W, Konieczny K *et al* 1998 Dielectric properties of $Li_{0.003}Na_{0.997}NbO_3$ ceramics *Ferroelectrics* **216** pp 53-61
[7] Reznichenko L A, Verbenko I A, Andryushin K P *et al* 2010 Lead-free long ago *Piezoceramic Materials and Devices* pp 1-69
[8] Hayakawa S I *et al* 1975 Electronic Ceramics *The proceedings of the symposium on electrical and electronic technology at the exhibition of electronic equipment* **56**
[9] Turik A V, Yesis A A , Reznitchenko L A *et al* 2006 Negative longitudinal electrostriction in polycrystalline ferroelectrics: A nonlinear approach *Journal of Physics Condensed Matter* **18** pp 4839–4843

IOP Conf. Series: Materials Science and Engineering 1093 (2021) 012021 doi:10.1088/1757-899X/1093/1/012021

Features of protolytic reactions in the ground and exited states in the presence of cationic polyelectrolyte

A S Mugabutaeva, A O Naumova, P V Melnikov and N K Zaytsev

Lomonosov Institute of fine chemical technologies, MIREA – Russian technological university, 86 Vernadsky Avenue, Moscow, Russia, 119454

Email: mugabutaeva@yandex.ru

Abstract. Protolytic reactions in the ground and excited states for 2-naphthol and its sulfo derivatives were studied. It was revealed that in the presence of cationic polyelectrolyte, the acidity constant of the protolytic reaction in the ground state shifts to a more acidic region. It was also found that the value of the shift depends on the number of sulfo substituents in 2-naphthol derivatives. It reaches 1.2 units for the disodium salt of 2-naphthol-3,6,8-trisulfonic acid. At the same time, no significant effect of the polyelectrolyte on the acidity constant in the excited state pK_a* was observed, and the photoprotolytic reaction proceeds in the same way as for an unimmobilized substance.

1. Introduction

Acid-base indicators are usually organic dyes, the molecules of which contain acidic or basic functional groups. Most often, indicators are used in the form of aqueous or alcoholic solutions added to the analyzed sample. Indicator test strips are also known, which include an indicator or a mixture of indicators in an adsorbed state. It is possible to quickly determine the acid-base properties of solutions by means of such systems [1-4]. The development of fluorescent indicator materials for the determination of various analytes is an actively developing scientific field. The most popular is the creation of sensors for pH [5, 6], O_2 [7], CO_2 [7-10] and other physiologically important parameters [11-15]. The phenomenon of photoinduced proton transfer has the unique properties that make it possible to measure the acidity of the medium using the ratio of bands in the fluorescence spectrum. This allows one to create a sensor without introducing a reference indicator [16].

When some substances are excited to the lower singlet excited state, their acid-base properties can change dramatically. The ability to enter into proton exchange reactions with solvent or other substances during the lifetime of an electronically excited state may appear. The mechanism of such acid-base reactions has attracted the attention of a large number of scientists since the time of the fundamental works of Weller and Eigen [17-20]. Upon excitation, some compounds become stronger bases (for example, acridine), while others, such as hydroxy- and amino-substituted aromatic compounds, become stronger acids [21]. α- and β-naphthols with an OH-group acting as a proton donor are the most typical model photoacids among a large number of known substances [22-24]. Usually, this class of molecules has an aromatic fragment and demonstrates a significant increase in acidity upon photoexcitation, which leads to the decrease in pK_a* by 5 or more units compared to pK_a in the ground state. It leads to deprotonation not only in alkaline, but also in neutral and even in acidic media [25].

Earlier, the possibility of creating colorimetric pH sensor based on Congo Red immobilized on SiO_2 was shown [26]. To create a fluorescent sensor, it is also necessary to study the effect of PDDA on the properties of the indicator. In this work, the effect of the cationic polyelectrolyte, polydiallyldiethylammonium chloride (PDDA), on pK_a* and pK_a using 2-naphthol and its sulfo derivatives as an example.

2. Experimental

2.1. Reagents

2-naphthol (**N0**), sodium salt of 2-naphthol-6-sulfonic acid (**N1**), disodium salt of 2-naphthol-3,6-disulfonic acid (**N2**), disodium salt of 2-naphthol-3,6,8-trisulfonic acid (**N3**) (Vekton, Russia) were used. Polydiallyldimethylammonium chloride (PDDA) solutions of the required concentration were prepared from the 20% stock solution (Sigma-Aldrich, USA) by dilution with distilled water (State standart GOST 6709-72). Buffer solutions with desired pH value were prepared from the original dry reagents ($Na_2HPO_4 \times 12H_2O$, $K_2HPO_4 \times 3H_2O$, KH_2PO_4, LenReaktiv, Russia) by the volume-weight method. The solutions of indicators with the required concentration were also prepared from the original dry reagents by the volume method. All experiments were carried out at a temperature of 25±2 °C.

2.2. Measurement technique

The pH value of the aqueous solutions of naphthol and its derivatives was monitored by means of Expert-001 pH-meter (Econix-Expert, Russia) equipped with ESK-10601/7 combined pH electrode (Izmeritelnaya Tekhnika, Russia). The concentration of the indicator was 1.57×10^{-6} mol/dm^3 in all solutions. The pK_a* value was calculated by processing the fluorescence spectra of solutions measured on the Cary Eclipse fluorescence spectrophotometer (Agilent Technologies, USA). A special setup (figure 1) was used to determine the pK_a value. It consisted of the pH-meter and eScope Pro DP-M17 USB microscope (OiTEZ, Hong Kong) connected to a PC running in-house software. The array of LED's with a wavelength of 375 nm was used for the solution excitation, since only deprotonated form is excited under these conditions. The investigated solution was placed in the cuvette compartment, and the image was taken automatically after stabilization of the pH meter reading. A series of solutions in phosphate buffer with pH from 5.7 to 10.3 were prepared for each indicator.

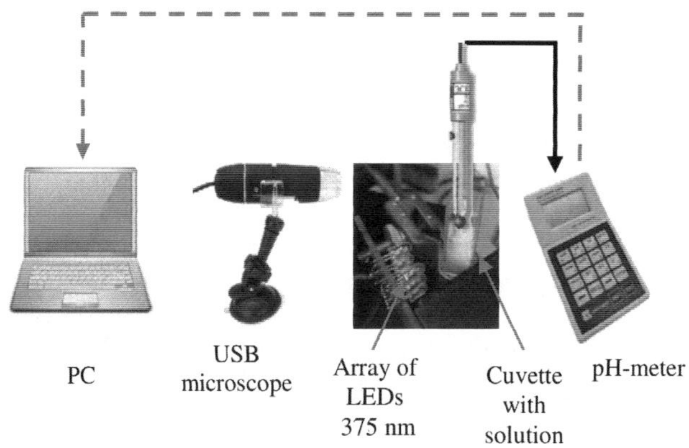

Figure 1. A Special setup for pK_a determination.

Each pixel in a digital image represents a set of three basic colors: red, green, and blue. The luminescence color of 2-naphthol and its sulfo derivatives varies from blue to blue-green, and the change in the intensity of both blue and green color channels can serve as an analytical signal. Such an assessment can be carried out for each pixel of the image independently. Therefore, in the future, it will be possible to assess local changes in pH.

3. Results and discussions

Protolytic reactions of 2-naphthol and its sulfo derivatives proceed according to the scheme [27, 28]:

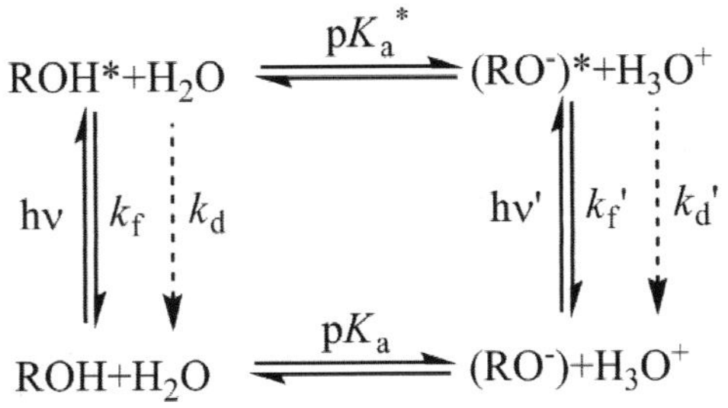

Dissociation occurs both in the ground (ROH) and in the excited (ROH*) state. The first process is characterized by the value of pK_a, and the second by pK_a*. In both cases, 2-naphthol and its derivatives are acting as donor molecules, and water molecules are acting as a proton acceptor.

First, a comparative study of solutions of 2-naphthol and its sulfo derivatives was carried out to determine the effect of the number of sulfo groups on their optical properties. Figure 2 shows the absorption and fluorescence spectra of the initial and deprotonated forms of **N1**. It can be seen that deprotonation leads to a shift of the peaks to a longer wavelength region, both in the absorption spectra and in the emission spectra. The initial form is observed in a strongly acidic medium, and the fully deprotonated form is observed in a strongly alkaline medium. A change in the pH of the medium leads to a change in the ratio of the intensities of the bands. It can be used as an analytical signal for pH determination without using an external reference dye.

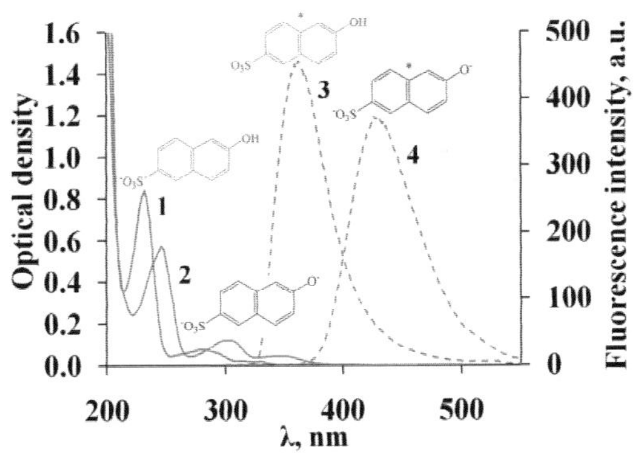

Figure 2. The absorption spectra of the initial (1) and deprotonated (2) forms of **N1** and the emission spectra of the initial (3) and deprotonated (4) forms of **N1**.

The dependence of the fluorescence spectra on pH for the example of **N1** and **N2** is shown in Figure 3a and b, respectively. It can be seen that the ratio of the amplitudes of the peaks of the initial (λ_{max} = 357 nm) and deprotonated (λ_{max} = 417 nm) forms strongly changes with pH change. Similar changes are observed for the fluorescence spectra of **N0** and **N3**. The increase in the number of sulfo groups in **N0** - **N3** series leads to a change in the intensity ratio of the observed peaks. As a result, the peak of the deprotonated form predominates in the **N3** fluorescence spectrum.

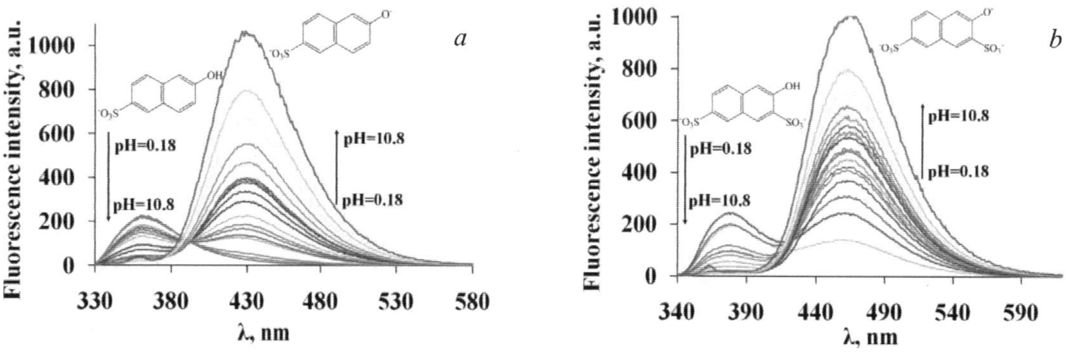

Figure 3. The dependence of the fluorescence spectra on pH for **N1** (a) and **N2** (b). C_{dye}= 1.57×10^{-6} mol/dm^3.

The change in the ratio of the intensities of the peaks of the initial and deprotonated forms of the indicator is shown by the example of **N2** in Figure 4. Two S-shaped sections are observed in the range of 0.5 ÷ 3.5 and 6 ÷ 11. The first corresponds to the photoinduced proton transfer reaction (pK_a*), and the second corresponds to the dissociation of the molecule in the ground state (pK_a). Therefore, the studied indicator systems can be used in both acidic and alkaline media.

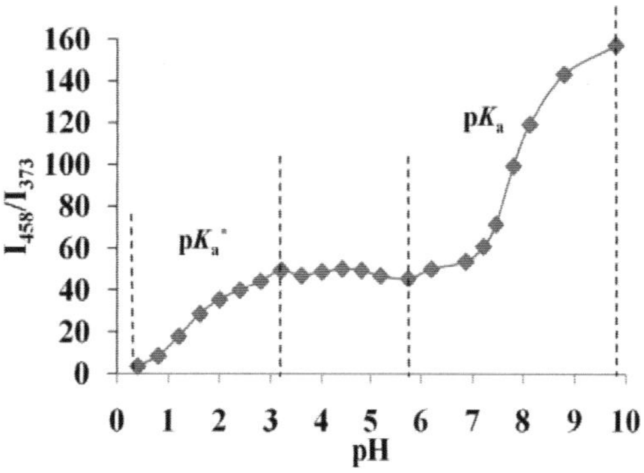

Figure 4. The change in the ratio of the intensities of the peaks of the initial and protonated forms (I_{463}/I_{377}) in the fluorescence spectrum of **N2** solution with a change in the pH of the medium.

Upon excitation with light λ_{max} = 375 nm, only the deprotonated forms of the molecules are excited. This makes it possible to study the dissociation process in the ground state, avoiding the influence of photoinduced proton transfer. The color intensity of the solution increases (goes from almost colorless to blue-green) with increasing pH. This is shown in Figure 5.

pH=4.8 pH=5.75 pH=6.20 pH=6.86 pH=7.67 pH=8.37 pH=9.0 pH=10.27

Figure 5. The changes in the fluorescence color of **N2** solution with increasing pH.

The effect of PDDA on the equilibrium constant of the deprotonation reaction in the ground state is shown in Table 1. It can be seen that the more SO_3^- substituents in the molecule, the stronger is the effect of the cationic polyelectrolyte on the system. This indicates a specific interaction of positively charged centers with negatively charged sulfo groups. Partial or complete compensation of the negative charge occurs, which facilitates the deprotonation of naphthol derivatives.

Table 1. The dependences of the pK_a and pK_a^* values for the indicators in distilled water and 0.05% PDDA solution.

Number of sulfo groups	pK_a			pK_a*		
	H_2O	PDDA	ΔpK_a	H_2O	PDDA	ΔpK_a*
0	9.45	9.43	-0.03	2.82	2.80	-0.02
1	8.94	8.29	-0.65	2.61	2.53	-0.09
2	8.57	7.52	-1.05	2.38	2.30	-0.08
3	8.13	6.93	-1.20	1.93	1.99	0.06

Despite the obvious influence of the cationic polyelectrolyte on the dissociation reaction in the ground state, which facilitates with an increase in the number of sulfo groups, the effect of the field formed by PDDA on the process of photoinduced proton transfer itself was not revealed. pK_a* values for all studied substances practically does not change when the polymer is introduced into the solution. The most probable reason for this is that upon photoexcitation, the indicator molecule becomes such a strong acid that the influence of the local environment can be practically neglected.

4. Conclusions

By the example of 2-naphthol and its sulfo derivatives, which are typical photoacids, it is shown that the addition of a cationic polyelectrolyte facilitates deprotonation in the ground state. In the considered series, the pK_a shift depends on the number of sulfo groups in the molecule. With the addition of PDDA, ΔpK_a monotonically increases from 0 for 2-naphthol (n = 0) to 1.2 for 2-naphthol-3,6,8-trisulfonic acid disodium salt (n = 3). Wherein, the local charge of the polyelectrolyte, which affects the dissociation process in the ground state, does not affect the reaction of photoinduced proton transfer. The luminescent properties of the indicator lower singlet-excited state do not change in this case. This allows the use of polyelectrolytes for the immobilization of photoacids or photobases to create flow sensors for pH.

Acknowledgments

The work was carried out using the equipment of the Center for Collective Use of Russian Technological University - MIREA with the support from the State Assignment of the Russian Federation in accordance with theme No. 0706-2020-0020.

References

[1] Zolotov Yu A, Ivanov V M, Amelin V G 2002 *Chemical test methods of analysis* (Moscow: Yeditorial URSS Publ) p 304
[2] Zolotov Yu A 1997 *Her. Russ. Acad. Sci.* **67 (6)** 499
[3] Amelin V G 2000 *J. Anal. Chem.* **55 (9)** 808–36
[4] Morozko S A and Ivanov V M 1995 *J. Anal. Chem.* **50 6** 629–36
[5] Staudinger C, Strobl M, Breininger J, Klimant I, Borisov S M 2019 *Sensor Actuat B-Chem* **282** 204
[6] Cai Q Y and Grimes C A 2001 *Sensor Actuat B-Chem* **79 (2–3)** 144
[7] Dixit R, Shen L, Ratterman M, Papautsky I, Klotzkin D 2012 *Proc. of SPIE* **8251** 825
[8] Oter O, Ertekin K, Derinkuyu S 2008 *Talanta* **76** 557
[9] Fritzsche E, Gruber P, Schutting S, Fischer J P, Strobl M, Müller J D, Borisov S M, Klimant I 2017 *Anal Methods* **9** 55
[10] Pfeifer D, Russegger A, Klimant I, Borisov S M 2020 *Sensor Actuat B-Chem* **304** 127312

[11] Fazial F F, Tan L L , Zubairi S I 2018 *Sensor Actuat B-Chem* **269** 36
[12] Zaragozá P, Fernández-Segovia I, Fuentes A, Vivancos J, Ros-Lis J V, Barat J M, Martínez-Máñez R 2014 *Sensor Actuat B-Chem* **195** 478
[13] Yang Z, Bai X, Ma S, Liu X, Zhao S, Yang Z 2017 *Anal Method* **9 (1)** 18
[14] Wang R, Diao L, Ren Q, Liu G, Pu S 2019 *ACS Omega* **4 (1)** 309
[15] Wu L, Sedgwick A. C, Sun X, Bull S D, He X-P, James T D 2019 *Acc Chem Res* **52** 2582
[16] Hynes J T, Klinman J P, Limbach H-H, Schowen R L 2007 *Hydrogen-Transfer Reactions* (Weinheim)
[17] Weller A 1955 *Naturwissenschaften* **42** 175
[18] Weller A 1958 *Zeitschrift fur Physikalische Chemie* **17 (3–4)** 224
[19] Weller A 1956 *Z. Elektrochem* **60 (9–10)** 1144
[20] Eigen M 1964 *Angew. Chem* **3 (1)** 1
[21] Wan P and Shukla D 1993 *Chem Rev* **93 (1)** 571
[22] Das A, Ayad S, Hanson K 2016 *Organic Letters* **18 (20)** 5416
[23] Li Y, Feng X, Wang A, Yang Y, Fei J, Sun B, Jia Y, Li J 2019 *Angew. Chem. Int. Ed.* **58** 796
[24] Cohen B, Martin Álvarez C, Alarcos Carmona N, Organero J A, Douhal A 2011 *J. Phys. Chem. B.* **115** 7637
[25] Simkovitch R, Shomer S, Gepshtein R, Huppert D 2015 *J. Phys. Chem. B.* **119 (6)** 2253
[26] Naumova A O, Melnikov P V, Dolganova E V, Yashtulov N A, Zaytsev N K 2020 *Fine Chem Tech* **15 (4)** 59–70
[27] Gerasimova M A, Tomilin F N, Malyara E Yu, Varganov S A, Fedorov D G, Ovchinnikov S G, Slyusareva E A 2020 *Dyes and Pigments* **173** 107851
[28] Tolbert L M, Solntsev K M 2002 *Accounts Chem Res* **35 (1)** 19

Surfacing of Rapidly-Wearing Parts

A S Nazarko and R L Plomodyalo

Federal State Budgetary Educational Institution of Higher Education "Kuban State Technological University", 2 Moskovskaya Street, Krasnodar, 350072, Russia

E-mail: nazarkoaleksandr@rambler.ru

Abstract. This paper considers the possibility of creating cast rods for surfacing of rapidly-wearing tillage tool parts of agricultural machinery, which would not contain scarce alloying elements and provide a wear-resistant deposited metal having high hardness and high resistance to crack formation in hardened condition. It was found that with an increase in boron concentration to 0.7 ... 0.8%, the amount of boride eutectic increases, which helps to prevent crystallization cracks, and the impact toughness first increases, and then slightly decreases, which is associated with the appearance of large boride inclusions.

1. Introduction

In surfacing rapidly-wearing tillage tool parts, most widely applied in practice are "Sormite No. 1" trademark electrodes containing a large number of scarce alloying elements - cobalt, tungsten, nickel, chromium, molybdenum, etc., which dramatically increases the manufacturing costs.

In solving the problem of reducing the amount of scarce alloying elements, as well as obtaining a wear-resistant deposited metal with high hardness and high resistance to crack formation in hardened condition, we proceeded from the following premises. Since the cooling rates of the deposited metal in the process of argon-arc surfacing and during air hardening are sufficiently close, when choosing a type of deposited metal, it is advisable to use air-hardening chromium steels of martensitic class as a basis. As far as is known, the hardenability of steels, i.e. their capability to increase their hardness, is mainly determined by their carbon content. To ensure the maximum hardness of martensite, the carbon content in it should be at least 0.5...0.6 % [1].

Chromium reduces the critical hardening rate, improves hardenability and forms special carbides, due to which the hardness and wear resistance of rapidly-wearing tillage tool articles increase. The chromium content in such steels can vary over a fairly wide range and reach 11 ... 13%.

To improve the hardenability and wear resistance of chromium steel, it is alloyed with manganese and silicon which are traditional deoxidizers of the weld pool when surfacing with cast rods [2]. The content of manganese in steels for the manufacture of tillage tools of agricultural machinery is usually 0.9...1.2 %; the amount of silicon can reach 3...5 %, however, the bending strength and impact toughness decrease.

Also of interest is the alloying of deposited metal with boron; it can be performed by using an affordably priced FB20 grade ferroboron introduced into the blend composition when smelting cast rods in a high-frequency induction furnace. Boron, even at low concentrations (up to 1%), significantly increases the hardness and wear resistance of deposited metal, but, as a rule, deteriorates its impact resistance [3].

2. Experimental section

Using the methods of mathematical design, we studied the "composition-properties" relationship for the deposited metal containing 0.4 ... 0.8% carbon, 1.0 ... 2.0% silicon, 7.0 ... 11.0% chromium, 0.2 ... 1.0% boron, as well as about 0.5% manganese and 0.2% titanium (the latter was added to the composition of cast rods as a processing aid).

For obtaining static models as second-degree polynomials, a design of Type Na-4 was used [4]. The variability intervals of the concentrations of alloying elements are presented in Table 1.

Table 1. Variability intervals of the concentrations of alloying elements.

Factors	Content, %			
	C	Cr	Si	B
Base level	0.6	9.0	1.5	0.6
Variability interval	0.2	2.0	0.5	0.4
Upper level	0.8	11.0	2.0	1.0
Lower level	0.4	7.0	1.0	0.2

The number of deposited metal properties that determine the operational reliability of tillage tools of agricultural machinery should include hardness, resistance to abrasive wear, bending strength, as well as heat resistance (under severe conditions, the tillage tools of agricultural machinery can be heated up to 200...300 °C).

Wearing tests of the deposited metal were carried out according to the well-known technique on a machine SMT-1. As a counterbody, specimens made of steel 45, hardened to 50 HRC, were used. The wear characteristic was the mass loss G_{wear}.

The bending strength $\sigma_{bending}$ was determined on specimens of 30x6x6 mm in size with a surface finish of $R_a = 0.4$ µm. To exclude errors introduced by friction forces, we used a device with roller bearings, the distance between which was 20 mm. The loading rate was 100 N/sec.

The characteristic of the heat resistance T_{50} of the deposited metal was a temperature of two-hour tempering after which the hardness decreased to 50 HRC.

One of the most common and critical defects encountered in surfacing alloy steels of martensitic class are cold cracks. The metallographic studies performed during preliminary experiments showed that in most cases cold cracks developed from small solidifying cracks. Therefore, when studying the "composition-properties" relationships, the resistance of deposited metal to hot cracking was chosen as one of the optimization parameters. For the quantitative assessment, the Apblett-Pellini test was used [5, 6]. The composite specimens were made from plates of steel 45 with dimensions of 100x15x15 mm. The surfacing was carried out in a mode of $I_w = 120 ... 140$ A. The criterion for the resistance of deposited metal to hot cracking was the percentage of the cross-sectional area of the bead occupied by hot cracks (S_{cr}).

3. Results and discussion

The chemical composition and test results of the deposited metal are shown in Table 2.

After processing the test results using multivariate regression analysis methods with sequential exclusion of non-significant variables, the following regression equations were obtained:

$$\begin{aligned} HRC = 51.911 - 5\sigma893Si + 20.231\,B - 1.0723\,C \times Cr + 11.065\,C \times Si - \\ - 0.66663\,Cr \times B + 0.049980\,Cr^2 - 7.3584\,B^2; \end{aligned} \quad (1)$$

multiple correlation coefficient $R = 0.9585$
residual standard deviation $\sigma_0 = 0.48$
significance level of the regression coefficient $P \leq 0.006$

$$G_{wear} = -62.206 + 5.2435\,Cr + 160.71\,Si - 160.47\,B - 50.176\,C \cdot Si - 7.0038\,Cr \cdot Si + 80.714\,C \cdot B + 8.4493\,Cr \cdot B + 2.8692\,Si \cdot B - 22.746\,Si^2 + 38.920\,B^2 \; [mg/mm^2]; \quad (2)$$

$$R = 9.889;\ \sigma_0 = 0.7;\ P \leq 0.54;$$

$$T_{50} = 405.39 + 469.51\,C - 111.88\,Si + 139.32\,B - 32.882\,C \cdot Cr - 136.53\,C \cdot B + 0.98134\,Cr^2 + 41.853\,Si^2 \; [°C] \quad (3)$$

$$R = 0.9657;\ \sigma_0 = 8.1;\ P \leq 0.017;$$

$$\sigma_{bending} = 29311 - 27767\,C - 3917.5\,Cr - 2544.7\,B + 1339.4\,C \cdot Cr + 7154.2\,C \cdot Si - 482.11\,Cr \cdot Si + 5275.1\,C \cdot B + 199.46\,Cr^2 \; [MPa]; \quad (4)$$

$$R = 0.9053;\ \sigma_0 = 35.5;\ P \leq 0.049;$$

$$S_{cr} = 667.44 - 368.28\,C - 72.378\,Cr - 447.92\,Si + 384.73\,B + 58.594\,C \cdot Cr + 42.630\,Cr \cdot Si - 306.18\,C \cdot B - 9.4500\,Cr \cdot B - 117.96\,Si \cdot B - 0.82950\,Cr^2 + 42.706\,Si^2 \; [\%]; \quad (5)$$

$$R = 0.9909;\ \sigma_0 = 1.8;\ P \leq 0.03;$$

Table 2. Chemical composition and test results of the deposited metal.

Alloy No.	Content, %				HRC	G_{wear}, mg/mm^2	T_{50}, °C	$\sigma_{bending}$, MPa	S_{cr}, %
	C	Cr	Si	B					
1	0.62	11.0	2.21	1.16	58	14.4	610	960	0
2	0.84	6.7	0.94	1.08	60	15.2	620	1020	0
3	0.32	11.0	0.96	1.17	57	21.6	600	2320	0
4	0.36	6.7	1.79	1.33	56	30.0	600	3280	0
5	0.80	11.2	0.91	0.25	54	13.2	550	840	27.2
6	0.75	7.7	1.84	0.24	56	20.0	570	1180	14.6
7	0.41	12.1	1.70	0.25	54	20.0	530	1180	35.5
8	0.40	7.8	0.96	0.21	53	26.6	530	2960	32.3
9	0.83	10.0	1.40	0.62	58	14.4	560	1110	15.2
10	0.44	10.3	1.43	0.58	57	22.2	540	1920	3.3
11	0.56	12.8	1.41	0.54	57	16.2	560	3280	1.6
12	0.67	7.5	1.39	0.57	58	18.0	560	1290	0
13	0.50	9.3	1.42	1.08	57	24.8	560	1080	0
14	0.61	10.3	1.44	0.23	55	20.0	520	1100	37.9
15	0.57	9.3	1.92	0.59	57	16.0	560	1050	20.5
16	0.57	10.0	0.94	0.59	56	13.5	550	980	5.6
17	0.60	10.0	1.38	0.54	57	18.8	560	1080	10.2

The analysis of the influence of alloying elements on the properties of the deposited metal with a constant content of the other elements in it showed, in particular, that increasing the boron content in the studied range increases the resistance of the deposited metal to formation of hot cracks. The favorable effect of boron can be accounted for by the appearance of boride eutectic in the structure of the deposited metal. In the deposited metal of Type 60X8C2PT, it appears already at a boron content of 0.2%, but in small amounts – insufficient for "healing" the small breaks of metal formed during primary solidification. With an increase in boron concentration to 0.7 ... 0.8%, the amount of boride eutectic also increases, which helps to prevent solidification cracks. With increasing the boron content in the specified range, the impact resistance first increases, and then decreases slightly, which is apparently associated with the appearance of large inclusions of borides [7].

The search for the optimal composition of deposited metal was carried out using a generalized objective function taking into account the values of individual indicators, since in this case there are several optimization parameters (hardness, strength, wear resistance, etc.). The analysis of literature data and production experience has shown that the deposited metal of tillage tools of agricultural machinery should meet the following requirements: HRC $\geq$ 58, $\sigma_{bending} \geq$ 2000 MPa, $T_{50} >$ 500 °C, $G_{wear} \leq$ 20 mg/mm^2, $S_{cr} = 0$.

As a result of the search under the given constraints, the optimal chemical composition of deposited metal was determined with the following expected properties: HRC 58, $\sigma_{bending}$ = 2050 MPa, T_{50} = 560 °C, G_{wear} = 18 mg/mm^2, $S_{cr} = 0$.

Cast rods were developed to obtain the tailored composition of deposited metal.

The structure of the metal, deposited with the cast rods, is: martensite, carbides, and borides in the presence of a moderate amount of retained austenite (figure 1).

The results of testing the specimens of the Type 60X8C2PT metal deposited with the developed cast rods are given in Table 3. As can be seen, the metal deposited with the cast rods is significantly superior in properties to steel grades L53 and L65 deposited with "Sormite No. 1" electrodes.

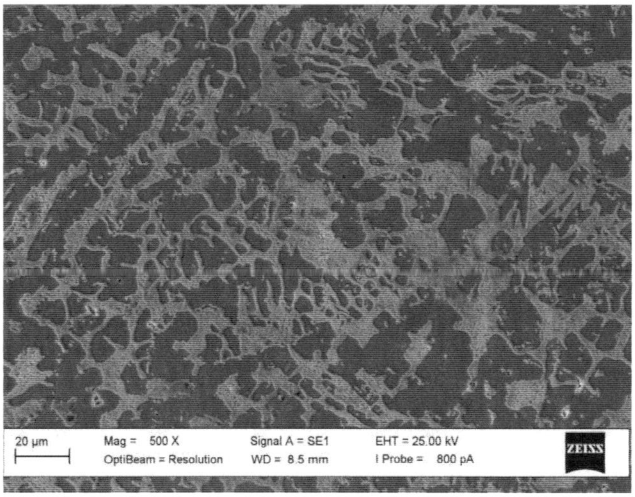

Figure 1. Microstructure of the metal deposited with the cast rods, x500.

Table 3. Test results for the deposited metal specimens.

Type of metal	HRC	G_{wear}, mg/mm^2	$\sigma_{bending}$, MPa	T_{50}, °C
60X8C2PT	58/59	18/19	2050/2000	560/550
L53	53	32	1050	570
L65	61	34	750	250

Note. The calculated data are in the numerator, the experimental data are in the denominator

Conclusion

Based on the experimental results obtained, it can be concluded that 60X8C2PT surfacing cast rods can be used both in the manufacture and in the repair of tillage tools of agricultural machinery.

The cast rods do not contain scarce alloying elements and provide a wear-resistant deposited metal having high hardness and high resistance to crack formation in hardened condition.

Acknowledgments
The research was carried out with the financial support of the Kuban science Foundation in the framework of the scientific project № МФИ-20.1/26².

References
[1] Dmitrenko V P, Manuilova N B 2017 Materials science in mechanical engineering, tutorial Moscow: INFRA-M p 432
[2] Nazarko A S, Plomodyalo R L and Demonov M S 2019 Composition for surfacing parts. Patent for invention RU 2705273, 07.11.2019. Application 2018147607 dated December 28, 2018
[3] Eremin E N, Losev A S and Borodikhin S A 2019 Structure and characteristics of the metal of the CR-NI-MO-MN-SI-TI-NB system obtained by surfacing with a flux-cored wire alloyed by boron carbide *Materials Science Forum* 946 MSF pp 91–96
[4] Sidnyaev N I 2017 Statistical analysis and theory of experiment planning: textbook *Moscow: Publishing house of MSTU* p 195
[5] Yakushin B F 2012 Comparative analysis of the ISO 18841 2005 standard and the current standard of the Russian Federation 26389-84 for the assessment of resistance to hot cracking during welding *Automatic welding* 11 (715) pp 46–49
[6] Nazarko A and Plomodyalo R 2019 Influence of chemical composition of wear-resistant hard-facing materials on the tendency to formation of hot cracks during weld facing *Materials Today: Proceedings* vol 19 part 5 pp 2422–2424
[7] Eremin E N, Losev A S, Borodikhin S A, Ponomarev I A, Matalasova A E 2019 Hardening of coatings based on chromium steel alloyed with a complex of boride compounds by quenching., Journal of Physics: Conference Series doi:10.1088/1742-6596/1210/1/012040

IOP Conf. Series: Materials Science and Engineering 1093 (2021) 012023 doi:10.1088/1757-899X/1093/1/012023

Using an Electromechanical Analogy to Describe the Damping Characteristics of an MFC Actuator

A A Pan'kov, A N Anoshkin, P V Pisarev and S R Bayandin

Perm National Research Polytechnic University, Perm, Russia

E-mail: krd2874@mail.ru

Abstract. In this article, a mathematical model was formulated that describes the dynamic behavior of MFC actuators. The model was developed on the basis of numerical-analytical methods for solving coupled boundary value problems of electromagnetoelasticity for inhomogeneous media equipped with piezoactive structural elements and loaded with coherent electromechanical disturbances. Within the framework of the performed numerical experiments, stationary harmonic axisymmetric oscillations of an elementary composite two-phase MFC actuator cell loaded with coherent electromechanical disturbances were considered. The calculation of the MFC actuators complex impedance and conductivity values, taking into account the energy losses and phase shift at different levels of the control electric voltage, was carried out.

1. Introduction

Highly loaded elements of aircraft structures, such as an aircraft wing or helicopter blades, create significant vibrations and noise due to changes in the aerodynamic loads acting on them when their azimuthal angle changes [1]. With the advent of piezoactive materials, it became possible to develop efficient and cost-effective SMART systems to vibration control [2]. Piezo actuators integrated into the wing skin or rotor blades made of polymer composite materials create dynamic twisting and curvature of the structure, adapting it at any time to flight conditions [3]. The use of such SMART structures in aviation technology can significantly reduce vibrations, noise and improve structure performance and flight performance in general [4].

At present, some works are known that describe the theoretical foundations of creating SMART structures from polymer composite materials with controlled geometry; in addition, there are examples of creating laboratory samples and working models of such structures [5-7]. However, the scientific problem limiting development of SMART structures (implementation in the practice of engineering design) is the lack of scientifically grounded approaches and proven methods of their design. To solve this problem, it is necessary to carry out a set of computational and experimental studies aimed at developing new mathematical models of composite materials with piezoactive structural elements (piezocomposites) and developing methods for solving related boundary value problems of electromagnetic elasticity for inhomogeneous media with piezoactive structural elements.

2. Numerical model

In this research a mathematical model was formulated to calculate the damping characteristics of the MFC actuator. The model describes dynamic piezoelectroelastic behavior of an MFC actuator elementary composite two-phase cell loaded with coherent electromechanical disturbances (1).

$$\sigma = \tilde{\sigma} e^{i\omega t}, \; U_{\text{con}} = \tilde{U}_{con} e^{i\omega t} \tag{1}$$

In this case, the mechanical stress $\sigma \equiv \sigma^*_{zz}(t)$ was set on the outer end surface of the actuator, and the control electric voltage $U_{\text{con}}(t)$ was set on the electrodes. When formulating the assumptions of the mathematical model, it was assumed that the σ and $U_{\text{con}}(t)$ values were proportional to the $e^{i\omega t}$ exponential and had a certain phase angle $\Delta_{U\sigma}$. The electrical conductivities of the actuator phases were taken into account through the complex form of recording the dielectric constants of the phases $\lambda_f = \lambda'_f - i\gamma_f / \omega$, where the real parts of the dielectric constants λ'_f, the specific electric conductivity γ_f of the phases $f = \overline{1,2}$, the angular frequency ω, which leads, first, to complex values and dependences on the angular frequency ω of the electroelastic fields in the phases and the resulting actuator characteristics and, secondly, to the occurrence of dispersion and energy losses in the actuator, known as "Maxwell-Wagner relaxation" [8].

For a homogenized elementary composite two-phase cell of periodicity, the complex values of the "total" dielectric axial permittivity

$$\lambda^*_{zz} \equiv \lambda^{(\sigma)*}_{zz} = \lambda^{*'}_{zz} - i\gamma^*_{zz} / \omega \tag{2}$$

with the actual value of permeability $\lambda^{*'}_{zz}$ and conductivity γ^*_{zz}, which was calculated

$$\lambda^*_{zz} = \alpha_D l \tag{3}$$

by the values of the α_D coefficient (4), taking into account the equalities

$$\widehat{D}_z = \lambda^*_{zz} \widehat{E}_z, \; \widehat{E}_z = U_{\text{con}} / l, \tag{4}$$

where "macrointensity" of the electric field $\widehat{E}_z$ between the end cross-sections of the actuator, the distance between the sections l was the "working" length of the actuator cell. The resulting electric charge at the actuator end cross-sections

$$q = \widehat{D}_z S, \tag{5}$$

where the area of the end cross-sections S and the amperage (between the end sections or between the electrodes of the actuator)

$$I = \frac{dq}{dt} = S\frac{d\widehat{D}_z}{dt} = S\tilde{D}_z \frac{d}{dt} e^{i\omega t} = i\omega S \widehat{D}_z \tag{6}$$

or in the $I = \tilde{I} e^{i\omega t}$ form with amplitude

$$\tilde{I} = i\omega S \alpha_D \tilde{U}_{\text{con}} \tag{7}$$

Complex values of the total resistance (electrical impedance) $\tilde{z}$ and $\tilde{g}$ admittance (admittance)

$$\tilde{z} \equiv \frac{\tilde{U}_{\text{con}}}{\tilde{I}} = Z e^{i\delta_{\text{IU}}}, \; \tilde{g} \equiv 1/\tilde{z} = G e^{-i\delta_{\text{IU}}}, \tag{8}$$

where the phase shift δ_{IU} between the current amperage and the control electric voltage on the electrodes, the total resistance $Z = \bar{U}_{\text{con}} / \bar{I}$ and the conductivity $G = 1/Z$ in the axial direction of the elementary composite two-phase cell "piezoelectric / housing (epoxy resin)", which are calculated based on the expansion of the quantity

$$J \equiv i\omega S \alpha_D = Z^{-1} e^{-i\delta_{\text{IU}}}, \tag{9}$$

taking into account equality (7), where absolute values are $\bar{U}_{\text{con}}$, $\bar{I}$. Known solutions [9] for the $Z_{1,2}$ impedance and the $\text{tg}\delta_{1,2}$ tangent of the phase angle of the current amperage and U_{con} voltage of an alternating current electric circuit containing active resistance R and electric capacitance C, in case of the parallel

$$Z_1 = \frac{R}{\sqrt{1+(\omega RC)^2}}, \quad \mathrm{tg}\delta_1 = \omega RC \tag{10}$$

and consistent

$$Z_2 = \sqrt{R^2 + (\omega C)^{-2}}, \quad \mathrm{tg}\delta_2 = 1/(\omega RC) \tag{11}$$

connections, where $R = R(\omega)$ and $C = C(\omega)$ values are calculated

$$R = \frac{l}{\gamma_{zz}^{*} S}, \quad C = \frac{\lambda_{zz}^{*'} S}{l}, \tag{12}$$

by the values of dielectric constant $\lambda_{zz}^{*'} = \lambda_{zz}^{*'}(\omega)$ and conductivity $\gamma_{zz}^{*} = \gamma_{zz}^{*}(\omega)$ between the end cross-sections of the actuator.

Axial displacement u and electric charge q at the end cross-section of the actuator can be represented by expansions of the form

$$u = a_{u1}U_{\text{con}} + a_{u2}\sigma = \alpha_u \sigma, \quad q = a_{q1}U_{\text{con}} + a_{q2}\sigma = \alpha_q U_{\text{con}} \tag{13}$$

with amplitudes

$$\tilde{u} = a_{u1}\tilde{U}_{\text{con}} + a_{u2}\tilde{\sigma} = \alpha_u \tilde{\sigma}, \quad \tilde{q} = a_{q1}\tilde{U}_{\text{con}} + a_{q2}\tilde{\sigma} = \alpha_q \tilde{U}_{\text{con}} \tag{14}$$

where the proportionality coefficients

$$\alpha_u = \kappa a_{u1} + a_{u2}, \quad \alpha_q = a_{q1} + \kappa^{-1} a_{q2}, \tag{15}$$

complex values of the amplitudes of the control electric voltage $\tilde{U}_{\text{con}}$ and mechanical stress on the end cross-section of the actuator $\tilde{\sigma}$, while we have the actuator full mechanical rigidity $k = \alpha_u^{-1}$ and electric capacity $C = \alpha_q$. In expressions (15), the proportionality coefficient is used

$$\kappa \equiv \tilde{U}_{\text{con}} / \tilde{\sigma} = \bar{\kappa} e^{-i\Delta_{U\sigma}} \tag{16}$$

for the amplitudes of the control electric voltage $\tilde{U}_{con}$ and $\tilde{\sigma}$ mechanical load, taking into account the equalities

$$\tilde{U}_{\text{con}} = \bar{U}_{\text{con}} e^{-i\Delta_{U\sigma}}, \quad \tilde{\sigma} = \bar{\sigma} \tag{17}$$

and notation

$$\bar{\kappa} = \bar{U}_{\text{con}} / \bar{\sigma}, \tag{18}$$

where the absolute values $\bar{U}_{\text{con}}$, $\bar{\sigma}$, phase shift $\Delta_{U\sigma}$ between the control electric voltage U_{con} and the diagnosed mechanical stress σ can be realized and adjusted to the optimal value using an external electrical circuit [10-13].

For the considered axisymmetric case of stationary vibrations, we have axial displacements u and stresses σ at the end cross section of the actuator

$$\sigma = \bar{\sigma} e^{i\omega t}, \quad u = \bar{u} e^{i(\omega t - \delta_{u\sigma})}, \tag{19}$$

in this case, the complex coefficient of the total elastic compliance of the sensor

$$s \equiv \alpha_u = u / \sigma = \tilde{u} / \tilde{\sigma} = \zeta e^{-i\delta_{u\sigma}} \tag{20}$$

where $\delta_{u\sigma}$ is the phase shift, the absolute values of the compliance coefficient $\zeta = \bar{u} / \bar{\sigma}$ and amplitudes of stress $\bar{\sigma}$ and displacement $\bar{u}$ at the end cross-section of the actuator.

Axial forces $F = \sigma S$ on the end cross-section of the actuator perform work for a T period

$$A_\sigma = \int_0^T F' du' = \bar{\sigma}\bar{u}S \int_0^T \cos(\omega t) d\cos(\omega t - \delta_{u\sigma}) = \pi \bar{\sigma}\bar{u}S \sin\delta_{u\sigma}, \tag{21}$$

transformed into thermal energy dissipated by the actuator, where the real parts of the force F' and displacement u', the absolute values of the stress $\bar{\sigma}$ and displacement $\bar{u}$, taking into account

$$\int_0^T \cos(\omega t) d\cos(\omega t - \delta_{u\sigma}) = \pi \sin\delta_{u\sigma}.$$

The coefficient of dissipation by the actuator of the elastic energy supplied from the outside was written in the form

$$A_\sigma / A_{\sigma 0} = 2\pi \frac{\zeta}{\zeta_0} \sin\delta_{u\sigma}, \quad (22)$$

where the designations of work $A_{\sigma 0} = \frac{1}{2} F_0 u_0 = \frac{1}{2} S\zeta_0 \bar{\sigma}^2$, force $F_0 = S\bar{\sigma}$, displacement $u_0 = \zeta_0 \bar{\sigma}$ and compliance coefficient ζ_0 are used for the case of quasi-static mechanical loading and the absence of electrical conductivity of the actuator phases.

As a result, the average value of mechanical power over the $T = 2\pi / \omega$ period of oscillation

$$< P_\sigma >= A_\sigma / T = \frac{1}{2} \bar{\sigma}\bar{u}\omega S \sin\delta_{u\sigma} = \frac{1}{2} \bar{\sigma}^2 \zeta\omega S \sin\delta_{u\sigma}, \quad (23)$$

while the power of the electric current has an average value

$$< P_U >= \frac{1}{2} \bar{I}\bar{U}_{con} \cos\delta_{IU} = \frac{\bar{U}^2_{con}}{2Z} \cos\delta_{IU} \quad (24)$$

taking into account instantaneous values

$$P_U(t) = U'_{con}(t) I'(t) = \bar{U}_{con} \cos\phi_U \bar{I} \cos(\phi_U - \delta_{IU}),$$

where the real part U'_{con} and the $\phi_U = \omega t - \Delta_{U\sigma}$ phase of the U_{con} control electric voltage, the real part of the current I'.

3. Results of numerical simulation

To calculate the impedance, a numerical simulation of a two-phase cell with the volume fraction of transversely isotropic fibers $v_1 = 0.53$ made of PZT-4 piezoceramic in an isotropic epoxy matrix with Young's modulus $E = 2.5$GPa and Poisson's ratio $\nu = 0.42$ was carried out. Figures 1 and 2 show the results of numerical simulation of the frequency dependences of the real $\mathrm{Re}[\lambda_{zz}^{(\sigma)*}]$ and imaginary $\mathrm{Im}[\lambda_{zz}^{(\sigma)*}]$ components of the $\lambda_{zz}^{(\sigma)*}$ dielectric constant of the actuator for the case of given amplitudes $\bar{\sigma} = 1$MPa, $\bar{U}_{con} = 1$V, phase displacement $\Delta_{U\sigma} = 0$, relative longitudinal dielectric constant of piezo fibers $\lambda_{PZT} / \lambda_0 = 635$, and vacuum permeability λ_0. Figure 2, b shows the results of calculating the impedance Z (○) and the cotangent of the phase angle $\mathrm{ctg}\delta_{IU}$ (○) (Figure 2, c) of the alternating current electric circuit in the sensor in comparison with approximate solutions for parallel Z_1 (△) (Figure 2, a) connection of frequency-dependent elements: R(ω) active resistance and C(ω) electrical capacitance (Figure 2 , b, c) depending on the ω angular frequency, obtained on the basis of known solutions [14]. Tangent of the angle of mechanical losses is shown in Table 1. For the case $\Delta_{U\sigma} = 0$ the electrical conductivity $\gamma = 5.3\mathrm{n}(\Omega \cdot \mathrm{m})^{-1}$, active resistance $\mathrm{R} = 43.55\mathrm{G}\Omega$ don't depend on ω frequency, while the $\mathrm{C} = 1.25$nF capacitance at $\bar{U}_{con} = 1$V, $\mathrm{C} = 0.038$nF at $\bar{U}_{con} = 100$V; the dependence of the R(ω), C(ω) quantities, on the ω frequency is observed in the presence of a phase shift $\Delta_{U\sigma} \neq 0$ (Figure 3).

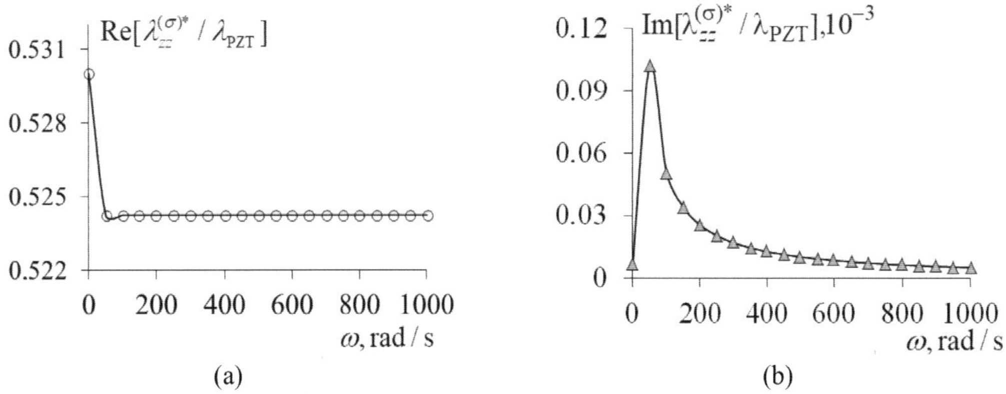

Figure 1. Real $\mathrm{Re}[\lambda_{zz}^{(\sigma)*}]$ and imaginary $\mathrm{Im}[\lambda_{zz}^{(\sigma)*}]$ components of the dielectric constant $\lambda_{zz}^{(\sigma)*}$ depending on the angular frequency ω at amplitudes $\bar{\sigma}$=1MPa, $\bar{U}_{con}$=1V, phase shift $\Delta_{U\sigma}$=0.

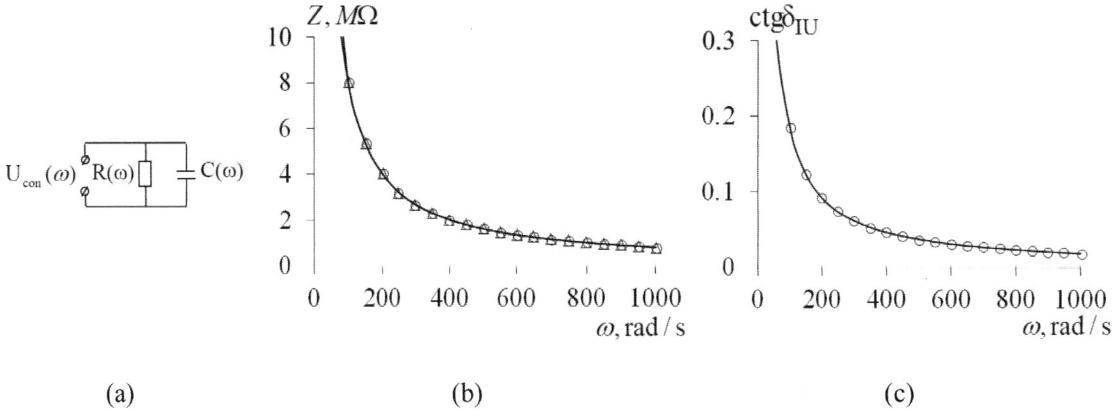

Figure 2. Impedance Z and cotangent of the phase angle $\mathrm{ctg}\delta_{\mathrm{IU}}$ depending on the angular frequency ω at amplitudes $\bar{\sigma}$=1MPa, $\bar{U}_{con}$=1V phase displacement $\Delta_{U\sigma}$=0 in comparison with the approximation by electromechanical analogy (a).

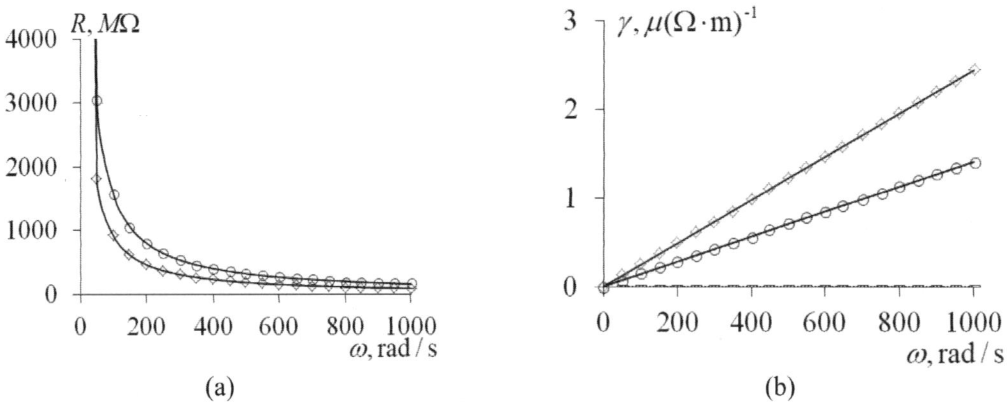

Figure 3. Active resistance R (a), electrical conductivity γ (b) depending on the ω angular frequency at amplitudes $\bar{\sigma}$=1MPa, $\bar{U}_{con}$=100V, phase shift $\Delta_{U\sigma}$=0 (▭), $\pi/6$ (○), $\pi/3$ (◊).

Table 1. Tangent of the angle of mechanical losses $tg\delta_{u\sigma}$.

$\Delta_{U\sigma}$	0	$\pm\pi/18$	$\pm\pi/6$	$\pm\pi/4$	$\pm\pi/3$	$\pm\pi/2$
$tg\delta_{u\sigma}$	0		± 0.226		± 0.573	

4. Conclusion

Thus, in this work, a mathematical model was formulated that describes the dynamic behavior of MFC actuators. The model was developed on the basis of numerical-analytical methods for solving coupled boundary value problems of electromagnetoelasticity for inhomogeneous media equipped with piezoactive structural elements and loaded with coherent electromechanical disturbances. Stationary harmonic axisymmetric oscillations of an elementary composite two-phase MFC actuator cell loaded with coherent electromechanical disturbances were considered within the framework of the performed computational experiments. The calculation of the complex values of the MFC actuators impedance and conductivity, taking into account the energy losses and phase shift at different levels of the control electric voltage, was carried out.

Acknowledgments

The research was carried out at the Perm National Research Polytechnic University with the support of the Russian Science Foundation (project No. 18-19-00722).

References

[1] Kovalovs A, Barkanov E and Gluhihs S 2007 *J. Phys. Conf. Ser.* **93** 012034
[2] Pisarev P V, Anoshkin A N and Ermakov D A 2019 *Mater. Today* **11** 446-452
[3] Barkanov E, Anoshkin A N, Pisarev P V and Ashihmin V A 2019 IOP C. Ser. – Mat. Sci. **511** 012029
[4] Anoshkin A N, Pisarev P V, Bayandin S R and Kungurtseva E G 2020 *AIP Conf. Proc.* **2216** 1
[5] Roth D and Enenkl B. 2006 *32nd European Rotorcraft Forum, Maastricht, the Netherlands*
[6] Janker P and Hermle F 2006 *32nd European Rotorcraft Forum, Maastricht, the Netherlands*
[7] Grohmann B A, Maucher C K 2008 *Mech. Adv. Mater. Struc.* **15** 307–324
[8] Petrov V M , Bichurin M I and Srinivasan G. 2004 *Tech. Phys. Lett+* **30** 8 81–87
[9] Dolginov A I 1957 *Resonance in electrical circuits and systems* (Moscow, Leningrad: Gosenergoizdat) p. 328
[10] Matveenko V P, Oshmarin D A, Sevodina N A and Yurlova N.A. 2016 *Computational mechanics of continuous media* **4** 476-485
[11] Kligman E P, Yurlov M A and Yurlova N A 2011 *PNRPU Mech. Bull.* **1** 106-120
[12] Yurlov M A, Oshmarin D A, Sevodina N V and Yurlova N A 2018 *PNRPU Mech. Bull.* **4** 266-277
[13] Belokon A V, Nasedkin A V and Danilenko A S 2007 *Bulletin of SamSU - Natural Science Series* **4** 56-65

IOP Conf. Series: Materials Science and Engineering 1093 (2021) 012024 doi:10.1088/1757-899X/1093/1/012024

Size Effects and Mechanisms in the Friction and Wear Processes of Ceramics Based on Zirconium Dioxide

T S Pirozhkova, A O Zhigachev, V V Korenkov and A I Tyurin

Nanotechnology and Nanomaterials Research Institute, G.R. Derzhavin Tambov State University, 33, Internatsionalnaya str., Tambov, 392000, Russia

E-mail: t-s-pir@ya.ru

Abstract. This paper presents a study of the tribological properties of ceramics based on ZrO_2 stabilized with CaO in a ceramic-ceramic friction pair. The influence of the scale factor on the coefficient of friction in the micro- and nanoscale is established. A qualitative correlation of the theoretical and experimental curves is shown for friction coefficient dependence on the body-counterbody contact force. Friction and wear micromechanisms at a micro- and nanoscale are proposed for the investigated TZP ceramics.

1. Introduction

Today rather strict requirements are imposed on modern engineering materials, especially on the ones used in high-load applications. Advanced nanostructured ceramics is one of the novel materials that satisfy these requirements. One of them is the engineering ceramic based on stabilized zirconium dioxide [1] that form an important class of materials with a unique set of properties, where strength and tribological characteristics come to the fore. Although the result of friction and wear is usually seen at the macrolevel, initially the interaction of interacting surfaces takes place at submicroscale and nanoscale through multiple local short-term contacts. As a rule, these contacts arise between the promontories of the surface due to the roughness; typical size of the contact regions ranges from a few nm to a few um [2-9]. It is a promising approach to simulate physically the processes occurring during dry friction at the level of single micro- / nanocontacts. This approach may allow us to get more detailed understanding and description of this process. It lets us to characterize the deformation processes occurring in locally loaded submicrovolumes, under fully controlled conditions.

2. Methods

The technique, used in this study, makes it possible to study the behavior of the material at local lateral interaction during sliding into the technique of nanoindentation [10-14], consists in measuring of local deformation with nanometer resolution. The local deformation occurs under the action of normal and lateral forces applied to the test material surface according to the test protocol [8-15]. The technique allows us to control of all essential parameters during the experiment (normal and lateral components of the forces, contact area and geometry of interacting areas of interacting bodies, local deformation and the lost material volume, and fracture toughness). It leads to creating completely defined conditions in dynamic micro- / nanocontact, vary them as desired, and investigate the mechanical behavior of materials on a nanoscale. The tribomechanical studies were carried out on a TI 950 Hysitron Triboindenter (USA) and DNT Nanoindentometers developed in the Research Institute

"Nanotechnology and Nanomaterials" branch of G.R. Derzhavin Tambov State University. Continuous recording of normal and lateral loading, as well as normal and lateral displacement, made it possible to evaluate the behavior of the material under the pressure of the tip touching the surface of the sample in fully controlled conditions at dependence on the displacement (Figure 1).

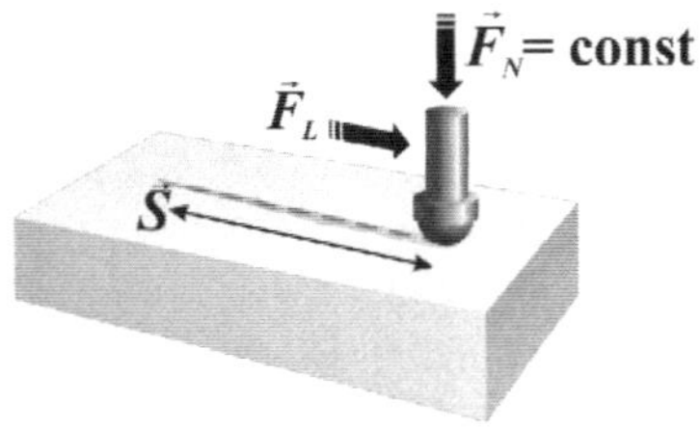

Figure 1. Schematic representation of friction and wear tests used in the present investigation.

The study shows tribological behavior of ZrO_2 stabilized with calcium oxide and ceramics with Nb_2O_5 additives were also used. A spherical tip made of ceramics based on ZrO_2 with a radius of R = 250 μm was used as a counterbody and the Berkovich indenter was used for small parts of tests.

3. Processes at micro- and nanoscale

Conducted research at the micro- and nanoscale has shown that the friction coefficient depends on the value of the normal force F_N under the tip (Figure 2). At the first stage, the friction coefficient was instantiated by quite a sharp decrease, and then a slight increase, and had a minimum at achieving a certain critical value of the normal force F_{Ncr}. Such behavior was represented typically for each set of tests and interacting bodies. However the F_N value varied and depended on the materials of body and counterbody. The plot until critical load F_{Ncr} corresponds to elastic interaction, which was also confirmed by the dependences of the load on the counterbody penetration depth $P = f(h)$ by reorganizing the obtained data during friction curves into P-h curves (Figure 3) where the load branch has the same slope like the unload curve. It is also confirmed by the absence of the tracks on the surface verified by atomic force microscopy.

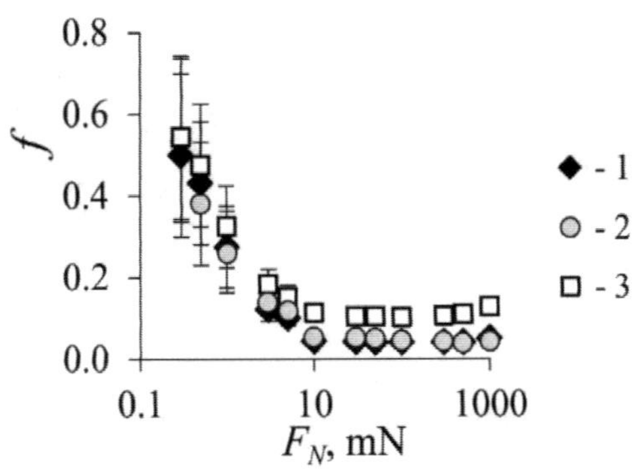

Figure 2. Dependence of the friction coefficient f on the normal load F_N: for the ZrO_2ceramics stabilised by 6.5 mol%. CaO and additives with 1 – 0%, 2 – 1,5 %, 3 – 5% Nb_2O_5.

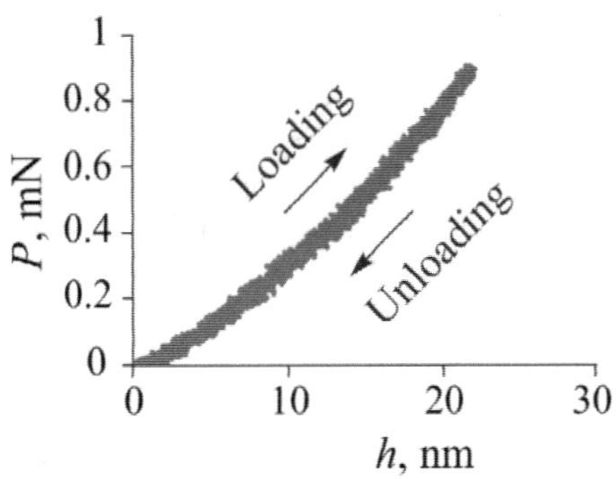

Figure 3. The typical load-displacement curve (P-h curve) for elastic zone ceramic reorganized from one of the friction curves.

Such authors as F. Bowden, D. Tabor, B.V. Derjaguin et al. described the friction process as the interaction of the surface irregular spots of contacting bodies what always occurs discretely and separately and bases on mechanical and molecular contribution components [9, 15-21]. The expansion of the friction coefficient obtained in our experiments according to the equations from the theory of the authors mentioned above leads to the fact that the total value of the friction force F_{fr} (numerically equal to the measured lateral component of the acting forces during friction F_L) is the sum of the molecular and deformation parts (F_{mol} and F_{def}), depending differently on the normal load F_N [15,16], and the coefficient of friction is determined by the next equation $f = f_{mol} + f_{def}$ [15,17,18].

According to the equation and the classical approach to tribology, the friction coefficient contains both the molecular and deformation components. The ratio of the contribution to the friction coefficient depends on many factors, such as surface topology, the nature of the contacting bodies, etc. [22-24].

The adhesive part of the friction coefficient f_{mol} at nanoscale (Figure 4) is dominant and contribute the main part into the obtained friction coefficient that was true for all friction pairs that were studied. The deformation part of friction coefficient f_{def} plays an unsignificant role when the normal loads are not high, and opposite, f_{def} gives the main contribute at relatively high loads $F_N > F_{Ncr.}$

The contribution of the deformation part is arising from 0 to 80% at F_N = 1000 mN, but at the same time, the contribution of the adhesive component is falling from ≈100% to 20% (at F_N = 0.2 mN and F_N = 1000 mN respectively) for the ZrO_2 ceramic with 6.5 mol%. CaO.

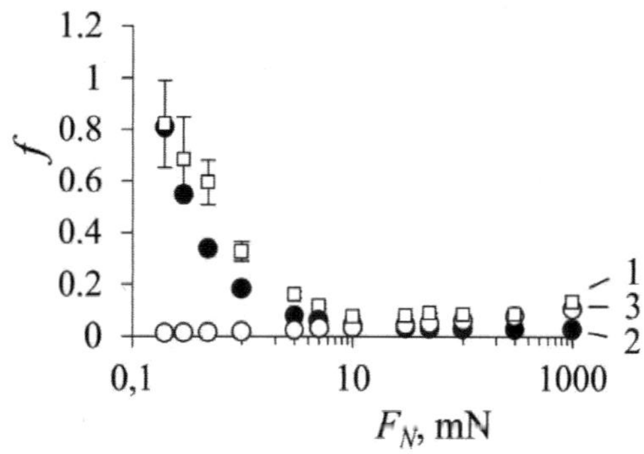

Figure 4. Interdependence of the friction coefficient f on the normal load F_N: for the ZrO_2ceramics with 6.5% CaO additives
1 – friction coefficient obtained in the tests; 2 – f_{mol} adhesive part of the friction coefficient; 3 – f_{def} deformation part of the friction coefficient.

4. Processes at macroscale

The carrying out of friction and wear at a macroscale were simulated by lateral sliding of a sharp tip (Berkovich indenter) under the impact of relatively high normal loads F_N (up to 1 N). This approach made it possible to form a sufficiently large deformation zone, where the processes of crack formation and destruction begin to be observed in addition to the typically observed track because of the counterbody sliding.

The results of experiments showed that separated cracks are detected under loads up to 0.2 N, further increasing of the load ($F_N \geq 0.5$ N) leads to the appearance of multiple cracking and spalling of ceramics separate grains. Figure 5. shows a typical example of the onset of cracking at dry friction of zirconium ceramics with 6.5 mol%. CaO.

A more detailed study of the identified cracks shows that they have a noticeable tendency towards a predominant direction of propagation, and it is to be about 60 degrees with the direction of the Berkovich's triangular indenter motion and corresponding to the highest tangential mechanical stresses existing along the ribs of the sliding tip.

The load increasing up to F_N = 500 mN leads to the fact that except the typical cracks, areas of greater destruction of the material were revealed inside the track left by the counterbody, in which spalling of individual grains was observed. Such spalling is characterized by a change in spatial orientation and practical exfoliation of individual grains.

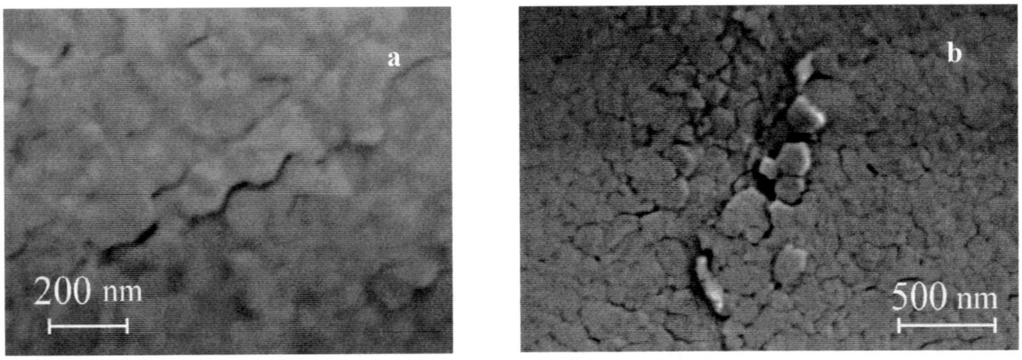

Figure 5. SEM of the track left by the Berkovich indenter on the surface of ZrO_2-based ceramics stabilized by 6.5 mol%. CaO: $F_N = 0{,}2$ N (a); $F_N = 1$ N (b).

The further increasing of the load When $F_N \geq 1$ N leads to the increasing in crack formation, and microerrosion of the surface is observed with chipping of fine particles like individual grains or their intergrowths and the formation of the debris.
The formation of cracks and further spalling of the material can be caused by the generation, displacement and further coalescence of point defects (for example, vacancies), destruction of grain boundaries, as well as phase transitions in the contact zone of the counterbody, which were revealed by micro-Raman spectroscopy.

5. Conclusions
It can be concluded that, purely elastic deformation mechanisms are initially implemented under the experiment conditions of dry friction, which are replaced by elastoplastic mechanisms with an increase of the normal load. When the load F_N keeps growing, crack formation and spalling of individual grains are added to elastic and elastoplastic mechanisms of deformation.

Based on this, we can say that the wear of the studied ceramics is based by a decrease in roughness at low loads (by cutting off singular protrusions of nano size), then mechanisms of plastic deformation begin to turn on with increasing in the load, and at the end, cracking and micro-erosion of the material are added at high loads, due to cutting off of the singular grains and restructuring of intergrowths borders.

Acknowledgments
The study was supported by a grant for young scientists of G.R. Derzhavin Tambov State University, order No. 267 of May 22, 2020

References
[1] Zhigachev A O, Umrikhin A V, Korenkov V V, Tyurin A I, Rodaev V V and Dyachek T A 2018 *Ceramic materials based on zirconium dioxide* (in Russian) ed Yu.I. Golovin, (Moscow: TECHNOSPHERE)
[2] Hu J and Ogletree D F 1995 Salmeron M. Atomic scale friction and wear of mica *Surface science*, **3** 327 358-70
[3] Carpick R W, Agrait N, Ogletree D F and Salmeron M 1996 Measurement of interfacial shear (friction) with an ultrahigh vacuum atomic force microscope *Journal of Vacuum Science & Technology B: Microelectronics and Nanometer Structures Processing, Measurement, and Phenomena* **14** 2 1289-95
[4] Zhang W et al 2019 Influence of surface roughness parameters and surface morphology on friction performance of ceramics *Journal of the Ceramic Society of Japan* **127** 11 837-42
[5] Yang M et al 2019 Effect of friction coefficient on chip thickness models in ductile-regime

grinding of zirconia ceramics *The International Journal of Advanced Manufacturing Technology* **102** 5-8 2617-32
[6] Basu B and Kalin M 2011 *Tribology of Ceramics and Composites* (New Jersey: Wiley & Sons, Inc., Hoboken)
[7] Bhushan B 2013 *Introduction to tribology* (New York: John Wiley & Sons)
[8] Bhushan B 1995 *Handbook of Micro/Nano Tribology* (New York: CRC Press LLC)
[9] Braunovic M, Myshkin N K and Konchits V V 2006 *Electrical contacts: fundamentals, applications and technology.* (London: Taylor &Francis Group, LLC)
[10] Oliver W C and Pharr G M 2004 Measurement of hardness and elastic modulus by instrumented indentation: Advances in understanding and refinements to methodology *Journal of materials research* **19** 1 3-20
[11] Golovin Yu I, Tyurin A I, Aslanyan E G, Pirozhkova T S and Vorobev M O 2016 Local physicomechanical properties of materials for use in calibration of nanoindentation instruments *Measurement Techniques* **59** 9 911-15
[12] Golovin Yu I, Tyurin A I, Aslanyan E G, Pirozhkova T S and Vasyukov V M 2017 The physical and mechanical properties and local deformation micromechanisms in materials with different dependence of hardness on the depth of print *Physics of the Solid State* **59** 9 1803-11
[13] Golovin Yu, I Ivolgin V I, Korenkov V V and Tyurin A I 1997 Determination of the time-dependent plastic properties of solids by dynamic nanoindentation *Technical Physics Letters* **23** 8 621-23
[14] Golovin Yu I, Tyurin A I and Iunin Yu L Strain-rate sensitivity of the hardness of crystalline materials under dynamic nanoindentation 2003 *Doklady Physics* **48** 9 505-08
[15] Burlakova V E, Belikova M A, Tyurin A I, Pirozhkova T S, Drogan E G, Novikova A A and Sadyrin E V 2019 Mechanical properties and size effects of self-organized film *Journal of Tribology* **141** 5 051601
[16] Myshkin N and Petrokovets M 2018 *Friction lubrication, wear. Physical foundations and technical applications of tribology* (in Russian) (Moskow: Litres)
[17] Tyurin A I, Pirozhkova T S and Shuvarin I A 2016 Research of processes of deformation when forming a print and friction in micro- and nanoscale (in Russian) *Izvestia of higher educational institutions. Physics* **59** 7-2 243-47
[18] Tyurin A I and Pirozhkova T S 2016 Research of the processes of friction and wear of solids in the micro- and nanoscale (in Russian) *Vestnik of Tambov University. Series: Natural and technical sciences* **21** 3 1375-80
[19] Bowden F P and Tabor D 1964 *Friction and lubrication of solids* (Oxford: Clarendon Press)
[20] Deryagin B V **1955** Problems of adhesion *Research (London)* **8** 70-74.
[21] Moscow, 1949 Kragelskii I V 1982 *Friction and wear* (Elmsford: Pergamon Press)
[22] Nakayama K, Bou-Said B and Ikeda H 1997 Triboelectromagnetic Phenomena of Hydrogenated Carbon Films -Tribo-electrons, -Ions, -Photons and -Charging, *J. Tribol.* **119** 764-68
[23] Hu J Ogletree D F 1995 Salmeron M Atomic scale friction and wear of mica *Surface science* **3** 327 358-70
[24] Carpick R W, Agrait N, Ogletree D F and Salmeron M 1996 Measurement of interfacial shear (friction) with an ultrahigh vacuum atomic force microscope *Journal of Vacuum Science & Technology B: Microelectronics and Nanometer Structures Processing, Measurement, and Phenomena* **14** 2 1289-95

Formation of high-entropy alloy by methods of ion-plasma technologies

N A Prokopenko, E A Petrikova, V V Shugurov, M S Petykevith, Yu F Ivanov and V V Uglov[2]

Institute of High Current Electronics Siberian Branch Russian Academy of Sciences, 634055, Tomsk, Russia
[2]Belarusian State University, 220030, Minsk, Belarus

E-mail: nick08_phantom@mail.ru

Abstract. Multi-element equiatomic single-phase thermodynamically stable substitutional solid solutions (high-entropy alloys) have a unique combination of mechanical, tribological, physical and other properties. The purpose of this research is to develop an ion-plasma method for the formation of high-entropy alloys. A method for the formation of high-entropy alloys has been developed, which consists in the deposition of a multi-element metal plasma on a substrate in a vacuum. Plasma is formed by vacuum-arc plasma-assisted simultaneous independent sputtering of the cathodes of the selected elements. The modes allowing synthesizing thin Ti-Al-Nb-Zr films with a composition close to equiatomic were revealed. The presence of droplet fraction particles of various elemental composition in the films was found. It is shown that the phase composition of the films depends on the concentration of elements in the plasma.

1. Introduction

High-entropy alloys (HEAs) are multi-element equiatomic single-phase thermodynamically stable substitutional solid solutions based mainly on BCC or FCC crystal lattice, possessing a unique combination of mechanical, tribological, physical and other properties, which makes it possible to recommend their use under extreme conditions in various industries [1-3]. The disadvantages of HEAs include the results presented in [4], which shows that HEAs made of refractory elements have high density (high specific weight) and fragility. The existing high-temperature HEAs with low density do not have high strength at high temperatures, with the exception of the NbMoCrTiAl compound, which demonstrated a yield strength of 600 MPa at 1000 ° C, but this alloy turned out to be very brittle at room temperature [5]. One of the promising ways to improve the plastic properties of heat-resistant HEAs without losing their strength is, as shown in [6, 7], the nanostructuring of the material. Alloy compositions with high entropy are currently obtained mainly by vacuum arc melting and mechanical alloying. For the creation of heat-resistant alloys from refractory elements, the methods of powder metallurgy seem to be practically significant [4].

The aim of this research is to develop an ion-plasma method for the formation of HEAs synthesized in the form of thin films, which consists in the deposition in a vacuum of a multi-element metal plasma created by an electric arc plasma-assisted simultaneous independent sputtering of the cathodes of the selected elements.

2. Material and methods

The HEA film is deposited using the plasma-assisted arc deposition method on the QUINTA installation (is part of the UNIKUUM complex, in the catalog of unique scientific installations, http://ckp-rf.ru/usu/434216/). The experimental scheme is shown in figure 1.

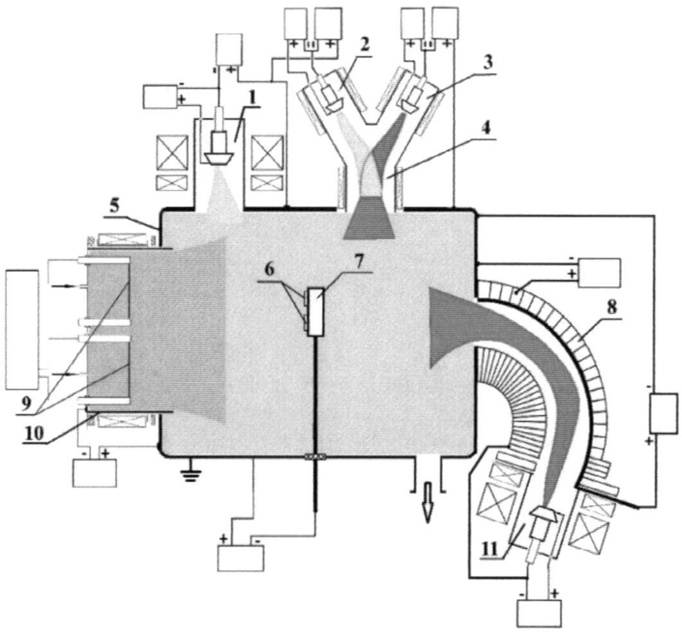

Figure 1. The scheme of the experiment for obtaining WPP:
1 – arc evaporator DI100 with Nb cathode, 2 – arc evaporator DI80 with Cu cathode,
3 – arc evaporator DI80 with Zr cathode, 4 – Y-shaped magnetic filter, 5 – vacuum chamber,
6 – samples, 7 – substrate holder, 8 – electromagnetic filter with a deflection angle of 120 degrees,
9 – heated cathodes, 10 – gas plasma generator PINK-P,
11 – arc evaporator DI100 with Ti-Al or Ti-Al-Si cathode.

For depositing, cathodes of copper M3, zirconium, niobium are used, as well as a composite cathode of composition (Ti-Al 50/50) or (Ti-Al-Si 45/45/10). Zirconium and copper cathodes are installed on a Y-shaped filter with two separate cathode units, a niobium cathode - on an arc evaporator DI100, and a composite Ti-Al or Ti-Al-Si cathode is installed on a magnetic filter with a plasma flow angle of 120 degrees. Deposition is carried out on substrates made of VK8 hard alloy.

After installing the samples (substrate) on the substrate holder, the vacuum chamber is pumped out to a pressure of $6.6 \cdot 10^{-3}$ Pa. Then argon is puffed up to a pressure of 0.3 Pa, the PINK-P gas plasma generator is switched on, and ion-plasma cleaning of the sample surface occurs. After cleaning, the metal plasma generators are started and starts the deposition of coating. After the end of deposition, the samples are cooled in a vacuum chamber to a temperature below 373 K.

The indisputable advantage of this method is the ecological cleanliness of the process, the possibility of complete automation of the ion-plasma deposition process and the formation of HEAs with a predetermined elemental composition, as well as the admissibility of the stage-by-stage formation of coatings with a strictly controlled elemental composition. This work presents the results of studying the structure and phase composition of four-elemental metal films, the thickness of which is 5 μm.

Films were sprayed onto WC-Co hard alloy. Investigations of the elemental and phase composition, the state of the defective substructure were carried out by scanning electron microscopy (Philips SEM-515 instrument with an EDAX ECON IV microanalyzer) and X-ray diffraction analysis (Shimadzu XRD 6000 diffractometer).

3. Results and discussion

Micro-X-ray spectral analysis was used to study the formed multi-element metal films. The results of the study, given in Table 1, indicate that the elemental composition of the obtained films can be controlled by changing the operating modes of the power supplies of the evaporated cathodes.

Table 1. Elemental composition of metal films.

Experiment	**Elemental composition, at. %**			
	Ti	*Al*	*Nb*	*Zr*
1	13.8	9.9	47.9	28.4
2	14.7	13.2	45.5	26.6
3	14.5	9.3	40.5	35.7
4	19.5	27.3	26.4	26.8
5	20.5	25.1	25.1	29.3

The surface structure of the deposited metal films has been studied by scanning electron microscopy (figure 2).

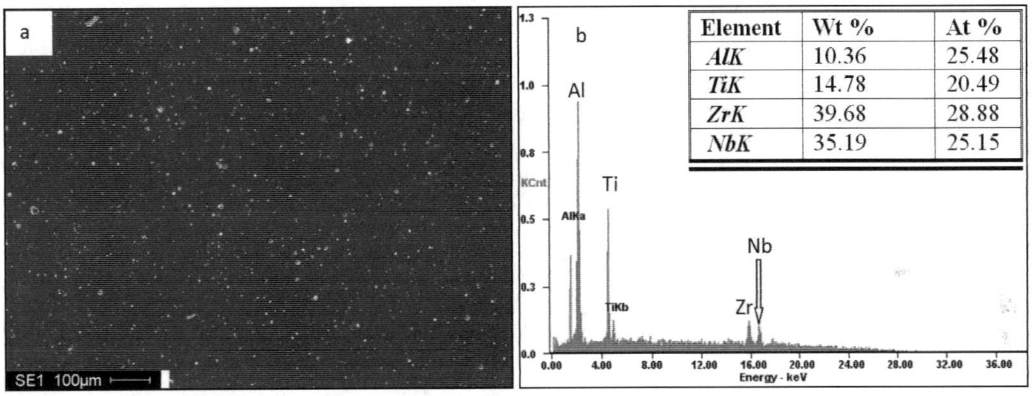

Figure 2. SEM image of the structure of films of the composition Ti-Al-Nb-Zr (a); (b) shows the energy spectra obtained from section (a) and the elemental composition of this section (table).

It was found that during the formation of films, irrespective of the concentration of the elements forming the film, particles of the microdroplet fraction are observed (figure 3). Particle sizes range from hundreds of nanometers to micrometer units.

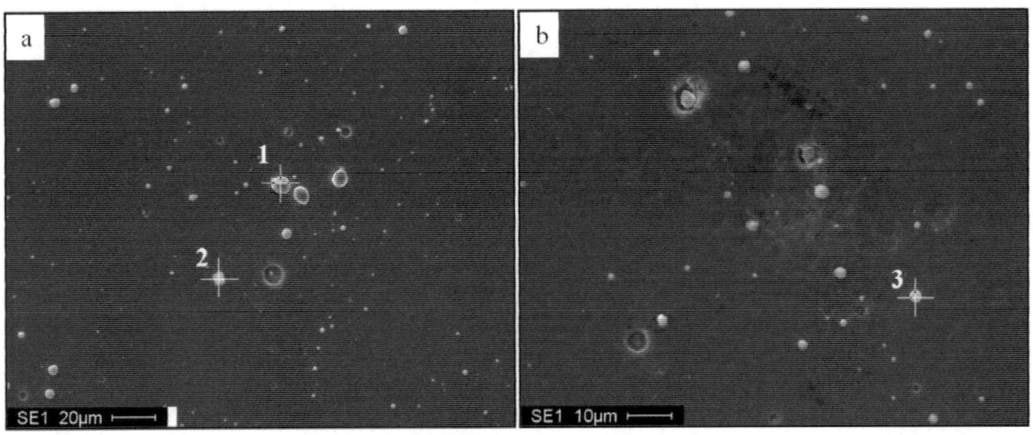

Figure 3. SEM image of the structure of films of the composition Ti-Al-Nb-Zr (a); numbers indicate drops, the elemental composition of which is given in table 2.

The elemental composition of the particles of the droplet fraction, indicated by numbers in figure 3, is shown in Table 2. It can be seen that the droplets can be enriched with atoms of different elements that form the film.

Table 2. Elemental composition of particles of micro-droplet fraction indicated in figure 3.

Area	Elemental composition, at. %			
	Ti	*Al*	*Nb*	*Zr*
1	21.6	37.6	22.1	18.7
2	59.6	16.5	12.2	11.7
3	7.8	15.8	66.5	9.9

The phase composition of the films was studied by X-ray structural analysis. It was found that the main phase forming the films is a solid solution based on Zr_xTi_y.

4. Conclusions

A method for the formation of high-entropy films has been developed, which consists in the deposition on a substrate in a vacuum of a multi-element metal plasma created by an electric arc plasma-assisted simultaneous independent evaporating of the cathodes of the selected elements. The possibility of controlling the elemental composition of the film by varying the operating mode of the arc discharge power supply sources is demonstrated. The modes are revealed that allow synthesizing thin Ti-Al-Nb-Zr films with a composition close to equiatomic. It is shown that the formed films contain droplet particles. It was found that the particles present in the films have different elemental compositions. It is shown that the phase composition of the films is a controllable characteristic and depends on the concentration of elements in the plasma, i.e. on the operating parameters of the evaporation system of the installation.

Acknowledgments

The study was supported by the Russian Foundation for Basic Research (project No. 20-58-00006 Bel_a).

References

[1] Ye Y F, Wang Q, Lu J, Liu C T and Yang Y 2016 *Materials Today* **19** 6 pp 349-362
[2] Pogrebnyak A D, Bagdasaryan A A, Yakushchenko I V, Beresnev V M 2014 *Uspekhi khimii* **83** 11 pp 1027-1061
[3] Miracle D B, Senkov O N 2017 *Acta Materialia* **122** pp 448-451
[4] Pickering E J, Jones N G 2016 *International Materials Reviews* **61** 3 pp 183-202
[5] Trofimenko N N, Efimochkin I Yu, Bolshakova A N 2018 *Aviacionnye Materialy i Tehnologii* 2 (51) pp 3-8
[6] Juan C C, Tsai M H, Tsai C W 2015 *Intermetallics* **62** pp 76-83
[7] Schuh B, Mendez-Martin F, Volker B, George E P, Clemens H, Pippan R, Hohenwarter A 2015 *Acta Materialia* **96** pp 258-268

Studies of water treatment influence on the structure of vein quartz from shungite rocks

Y L Rigaeva[1,2], N N Rozhkova[1] and T A Ekimova[2]

[1]Institute of Geology Karelian Research Center RAS, 11, Pushkinskaya str., Petrozavodsk, 185910, Russia
[2]Petrozavodsk State University, 33, Lenina ave., Petrozavodsk, 185910, Russia

E-mail: juliarigaeva@gmail.com, rozhkova@krc.karelia.ru, dery77@mail.com

Abstract. The article presents the results of the research of the structure and composition of one of the morphological varieties of quartz – vein quartz of shungite rocks from Zazhogino deposit of Karelia and a reference quartz sample. The values of the crystal lattice parameters were determined by full-profile analysis (Rietveld method). Coherent scattering region (CSR) for vein quartz samples are determined using the Debye-Scherrer formula after applying the background subtraction function. The influence of water treatment on the content of impurity elements in samples of vein quartz was also studied. The most common impurity elements in natural vein quartz samples, such as Al, Na, Li and Ge, are identified. It is shown that water treatment is effective for reducing the content of impurity elements and obtaining the minimum size of vein quartz crystallites.

1. Introduction

Recently, more and more attention is being paid to the possibilities of obtaining and applying nanoparticles in various production areas. Nano- and biomedicine occupy a special place here. It is these areas that are interested in the possibilities of developing and enhancing expanding the possibility of various applications of nanoparticles and nanomaterials [1, 2].

Quartz is a popular material in many industrial areas, such as machine and aircraft construction, production of new-generation composite materials, biomedicine, etc. Each application area requires used material to comply with strict parameters, such as dispersion and purity [3]. The size of a nanoparticle is a critical parameter that largely determines its properties. Depending on the size, the properties of nanoparticles, such as toxicity, adsorption properties, and many others, may change [4]. Therefore, the preparation of nanoparticles is a challenging task.

Quartz content is one of the main criteria for distinguishing of the most valuable industrial types of shungite rocks from the currently developed Zazhogino field. Quartz is represented by two genetic varieties that differ in the morphology of manifestation and time of formation. The first is terrigenous-chemogenic microsized and nanosized quartz of massive shungite rocks. The second is hydrothermal quartz presented as consisting of in veins and veinlets [3, 5, 6].

Regardless of the rock composition, quartz of all varieties is well crystallized and is characterized by nanosized crystallites [7].

The aim of this work is to study the structure of vein quartz of shungite rocks from Zazhogino deposit of Karelia and search for ways to clean it.

2. Materials and Methods

Quartz grains were selected manually for the study (figure 1 a). Then, the obtained grain sample was ground to powder (figure 1 b) with the particle size less than 40 microns (sample 1) and was subjected to purification. To remove soluble impurities, initial quartz powder was repeatedly washed with water to obtain an aqueous suspension of quartz particles (figure 1 c) and then filtered.

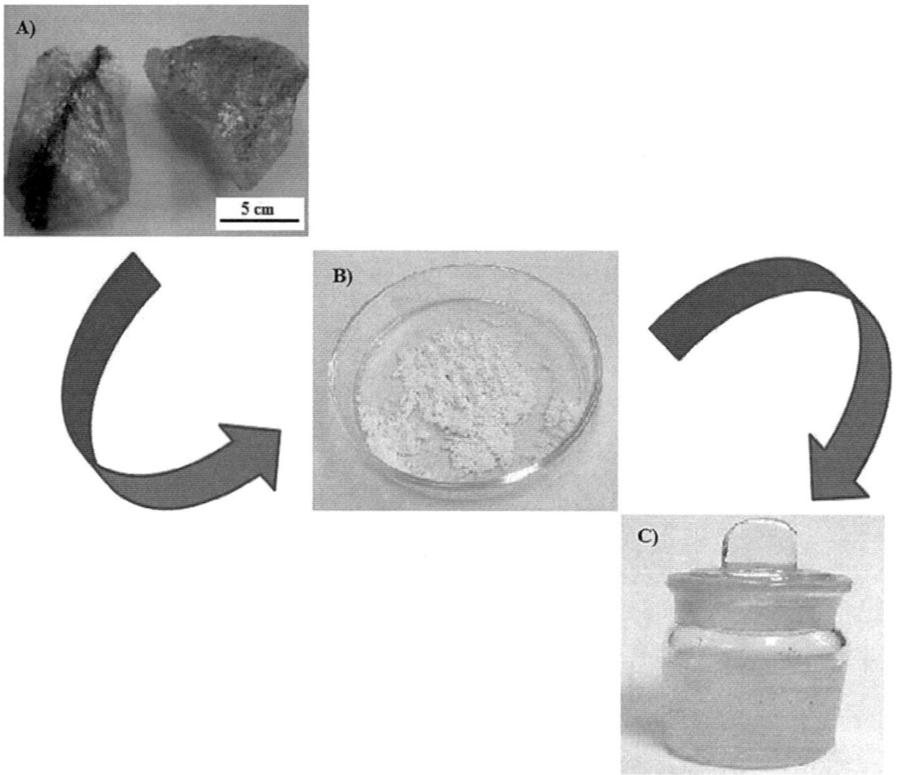

Figure 1. Scheme of obtaining the samples of vein quartz:
a) quartz grains from veins of shungite rocks; b) powder after water treatment (sample 2); c) suspension passed through the filter, sample 3 was obtained after drying (sample 3).

X-ray diffraction measurements of sample performed with the laboratory diffractometer DRON-6 with a monochromated CuK_α radiation source (λ = 1.54178 Å), 2θ range 19°-145.

Qualitative phase analysis and Rietveld full-profile analysis were performed using PDWin software [8].

The quartz coherent scattering region (CSR) was determined based on the maximum intensity peak of diffraction reflection (101) in the angular region 2θ from 26.54° to 26.66° according to the Debye-Sherrer formula after the application of background subtraction function:

$$D_{hkl} = \frac{0.9 \cdot \lambda}{b \cdot \cos\theta}, \quad (1)$$

where λ is the monochromatic wave length CuKα = 1.540562 Å, b is the peak width at half the height of the line, θ is the diffraction angle. The commonly accepted error in calculating the ACL with dimensions up to 200 nm is about 20% [9].

Using the reflection (101) which characterizes the sample as belonging to quartz, the quartz coherent scattering region was defined using the Debye-Sherrer formula (equation 1) that is assumed to be the average crystallite size in the studied sample.

The X-ray phase analysis showed that the X-ray patterns of all studied samples are qualitatively similar and have the distinctive appearance of crystalline α-quartz.

The Rietveld method was used to specify structural and profile characteristics of X-ray patterns of the studied samples. Initial data on crystallographic characteristics, coordinates, atomic vibration parameters and thermal parameters were taken from the ICSD database.

The Rietveld method includes the refinement of structural and profile parameters. At the first stage, profile parameters are refined, such as background polynomial coefficients, profile function parameters, periods and unit cell angles. At the second stage, structural parameters, such as atomic coordinates and atomic thermal vibration coefficients, are specified. Only after obtaining satisfactory results of the refinement of the structural and profile characteristics we may then proceed to introducing impure atoms in the samples into the refinement process.

Precision study of the content of impurity elements in vein quartz of shungite rocks of Zazhogino deposit of Karelia was performed using a Thermo scientific XSeries-2 ICP-MS quadrupole mass-spectrometer with the new Wave research UPP266 Macro laser ablation attachment (Nd:YAG laser, radiation wavelength 266 nm, pulse energy 0.133 MJ, scanning speed 70 microns/s, pulse repetition frequency 10 Hz, ablation spot diameter 515 microns). The measurement error on the mass spectrometer was taken as the value of the standard deviation S when determining the concentration of each element in the sample under study. The absolute measurement error, expressed in mg/kg, is equal to the value of the standard deviation, and the relative error (as a percentage) is equal to the relative standard deviation [12].

3. Results

The parameters of unit cell refined by the full-profile analysis as well as calculation reliability were estimated with the help of R-factors (Table 1). Profile (Rp) and weight-profile (Rwp) factors characterize the accuracy of agreement of the profile of the experimental and theoretical X-ray diffraction patterns in the entire range of dispersion angles (Rp) and mainly in the region of high intensity lines (Rwp), respectively. Bragg factors (Re and GofF) give an assessment of the crystal structure model.

The parameters of the quartz crystal lattice refined by the full-profile analysis method (Table 1) show that the removal of impurities from the studied samples affected the parameters of the unit cell.

Table 1. Structural parameters and criteria of values' adjustment obtained as a result of profile and structural parameters specification of vein quartz samples by the Rietveld method in PDWin software package.

Sample	a = b[Å]	c [Å]	α = β [°]	γ [°]	Rwp [%]	Rp [%]	Re [%]	GofF
Starting ICSD_174	4.9134(0)	5.4052(0)	90.0	120.0	-	-	-	-
1	4.9128(1)	5.4038(7)	90.0	120.0	17.16	12.23	12.89	1.33
2	4.9135(8)	5.4046(8)	90.0	120.0	12.02	9.23	13.01	0.92
3	4.9138(6)	5.4048(8)	90.0	120.0	9.14	7.10	6.40	1.43

All studied samples contain well-crystallized α-quartz as the obtained structural parameters indicate. The samples of vein quartz belong to low-temperature α-quartz since the vein quartz is characterized by trigonal lattice (they are related to P 3 2 2 1 space group) [13-15].

It is known that structural distortion can be influenced by impurity elements, adsorptive water, occurrence of OH-group instead of oxygen [16, 17]. ICP-MS is one of the basic methods of analysis of impurity content. Studies of the content of impurities in the vein quartz were carried out in Institute of Geology Karelian Research Center RAS.

Table 2. Content of impurity elements in quartz samples.

Sample	Li, ppb	Na, ppb	Mg, ppb	Al, ppb	Ti, ppb	Ge, ppb
1	0.381	218.4	58.63	6.851	88.43	0.027
2	0.321	18.63	0.73	-	0.182	0.015

The amount of structural impurities mainly depends on the genesis and thermodynamic conditions of rock formation [18]. The results presented in Table 2 show that the most common impurities are Al, Li, Na and Ge, the content of which varies depending on different treatments of the initial quartz sample and can be reduced with water treatment.

The coherent scattering region characterizes the average size of crystallites. The studied vein quartz of shungite rocks is hydrothermal quartz, so the main factor affecting the change in the parameters of the crystal lattice [17, 19] and CSR is the crystallization temperature. The CSR value is the most characterizing for determining the size of pure quartz crystallites [15-17].
In this paper, the calculation was performed using the Debye-Scherrer formula (equation (1)).
Results are presented in Table 3.

Table 3. Apparent coherent length of vein quartz samples.

Sample	1	2	3
ACL, nm	114.69	104.28	118.67

The results presented above (see Table 3) show that the CSR values in the direction (101) for all samples have a size close to 100 nm. However, sample 2 is highlighted by the lowest CSR value. According to the results of studies (see Table 2, Table 3), it can be said that water treatment of the sample leads to a decrease in the content of impurity elements, and, consequently, a decrease in the size of the crystallites and stabilization of the structure. Thus, quartz of the shungite rock from Zazhogino field of Karelia has a nanometric size.

4. Conclusion

The values of crystal lattice parameters of vein α-quartz sample of shungite rocks were determined by X-ray diffraction analysis using the Rietveld method.

The coherent scattering regions of vein α-quartz sample of shungite rocks were calculated using the Debye-Scherrer formula. Quartz sample 2, treated with water, is characterized by a minimum content of impurity elements. This type of processing provides a minimum size of crystallites of vein quartz of shungite rocks.

Acknowledgments

The work was carried out within the budget funding of Institute of Geology Karelian Research Center RAS AAAA-A18-118020690131-4.

References

[1] Roduner E 2010 Size effects in nanomaterials *Technosphere*
[2] Chang X L, Yang S T, Xing G 2014 Molecular toxicity of nanomaterials *J. Biomed. Nanotechnol.* **10** 2828-2851
[3] Sadovnichiy R V, Mikhaylina A A, Rozhkova N N, *et al 2016* Morphological and structural shungite quartz features of Maksovskiy deposit *Proceedings of the Karelian research center of the RAS Precambrian geology series* **2** 73-89
[4] Nikiforov V N, Filinova E Yu 2009 Biomedical applications of the magnetic nanoparticles *Magnetic Nanoparticles* **10** 393-455
[5] Frank-Kamenetskii V A and Kamentsev I E 1967 Microtonalism and formation conditions of

quartz *Problems of crystal chemistry of minerals and endogenous mineral formation* 68-76
[6] Komov L, Samoilovich M I 1985 *Natural quartz and its physical and chemical properties* Nedra
[7] Melnikov E P 1988 *Geology, genesis and industrial types of quartz deposits* Nedra
[8] Aleshina L A 2004 *Crystal Radiography: Study Guide* Petrozavodsk State University
[9] Kamentsev E 1970 The use of quartz as a reference for X-Ray studies and the accuracy of determining the parameters of its unit cell *X-ray analysis of mineral raw materials* **7** 108-115
[10] Sochalski-Kolbus L M 2015 Solvothermal synthesis and surface chemistry to control the size and morphology of nanoquartz *Crystal Growth & Design* **15** 5327-5331
[11] Krzhizhanovskaya M G, Firsova V A and Bubnova R S 2016 *Application of the Rietveld Method for Solving Powder Diffractometry Problems: Study Guide* Saint Petersburg State University
[12] Danilevskaya L A 2006 Quartz raw materials of Karelia region *Proceedings of Karelian Research Centre of RAS* **6** 50-66
[13] Svetova E N and Inina I S 2012 X-Ray structural features of industrial varieties of quartz hydrothermal metamorphogenic veins of the Polar Urals in: *Materials of the VI Russian Seminar on Technological Mineralogy* pp 102-105
[14] Kuznetsov S K, Lyutoev V P and Shanina S N *et al 2011* Features of quality of vein quartz of Ural deposits *News of the Komi Scientific Centre of the Ural Department of RAS* **4(8)** 65-72
[15] Rozhkova N N, Sadovnichy R V and Mikhailina A A *et al 2015* Structural homogeneity of quartz of schungite rocks *Petrography of magmatic and metamorphic rocks: Materials of the XII All-Russian petrographic meeting* 484-486
[16] Rozhkova N N, Rozhkov S S, Kovalchuk A A *et al 2016* Nanoscale carbon and quartz of shungite rocks *Nanostructural material – 2016: BELARUS – RUSSIA – UKRAINE: Materials of the V International Scientific Conference* 392-395
[17] Kuznetsov S K, Svetova E N, Shanina S N *et al 2012*, Hydrothermal-metamorphic quartz veins impurity elements of the Nether-Polar Urals region, *Geochemistry* **9** 1-16
[18] Kamentsev E 1962 Influence of crystallization conditions on the entry of foreign impurities into the quartz lattice *Herald of the Leningrad University Geology and geography series* **18** 109-112
[19] Lazarev D A and Kanygina O N 2012 On determining the size of regions of coherent scattering of quartz crystals in clay systems *Herald of Orenburg State University* **4(140)** 221-225

IOP Conf. Series: Materials Science and Engineering 1093 (2021) 012027 doi:10.1088/1757-899X/1093/1/012027

Dislocation structure and phase composition of Ti13Nb13Zr after high-pressure torsion

E A Rudenya[1], I V Ivanov[1], A Yu Ognev[1] and A M George Jr[2]

[1]Department of Material Science in Mechanical Engineering
Novosibirsk State Technical University, 20, Prospekt K. Marksa, Novosibirsk, Russia
[2]Materials Engineering Department, Federal University of San Carlos, Rod. Washington Luiz, s/n, São Carlos - SP, 13565-905, Brasil

E-mail: i.ivanov@corp.nstu.ru

Abstract. The approach based on synchrotron X-ray diffraction was applied in this research to investigate the dislocation and phase structures of severely deformed disk of Ti13Nb13Zr alloy. It was found, that the final structure of the material is inhomogeneous along the disk diameter. The local area of highest dislocation density of α'-phase matches with the zone of passed deformation-induced phase transformations.

1. Introduction

High-pressure torsion (HPT) is one of the severe plastic deformation (SPD) methods. It is widely used to study the behavior of metallic alloys under high pressure and strains. A characteristic feature of this technological process is different conditions of external loading of peripheral and central parts of the deformed material. For this reason, the structural transformations occurring along the diameter of the sample are different. This study describes the results of structural and phase analysis of martensitic titanium alloy Ti13Nb13Zr subjected to SPD by the HPT method. Several great properties of this alloy (e.g. the great mechanical resistance, low elasticity modulus and high corrosion resistance) make it a promising alternative to a classical biomedical titanium alloys (such as cp-Ti and Ti6Al4V) [1]. This research is based on a complex approach. It combines several deconvolution profile analysis methods to fully describe the transformations of micro- and dislocation structures of deformed by HPT material.

2. Experimental procedure

The material used in this study was the biomedical Ti13Nb13Zr titanium alloy. A disk with a thickness of 1 mm and a diameter of 10 mm was used for high-pressure torsion. Before performing deformation, it was annealed in a vacuum furnace for 1 hour at a temperature of 750^0 C and cooled in water. HPT was performed by using an equipment of Federal University of San Carlos, Brazil. The pressure during the experiment was 4.5 GPa. The upper anvil rotated 3 turns. The rotation speed was 3 rpm.

Scanning electron microscopy was performed by using a Phillips XL 30 FEG scanning microscope. Conventional transmission microscopy and orientation-phase mapping were accomplished by transmission electron microscopy (TEM) using a FEI TECNAI G2 operated at 200 KV. This facility is equipped with an orientation and phase mapping precession unit NanoMEGAS (model ASTAR) and Digistar P1000 unit.

The microhardness was evaluated using a WOLPERT Group 402 MVD Vickers hardness tester under a load of 100 g and a dwell time of 10 s.

Diffraction experiments were performed in transmission mode using synchrotron radiation on the P07 beamline (The High Energy Materials Science) at Deutsches Elektronen-Synchrotron (DESY) (Hamburg, Germany). The used wavelength was 0.014 nm. A 16-inch 2D PerkinElmer XRD 1621 scintillation detector with resolution of 2048 × 2048 pixels was used to record the diffraction patterns. The distance from the test material to the detector was 1.8 m. The images were taken along the diameter of the disk, the beam size was 1×1 mm.

Two-dimensional diffraction patterns were obtained in diffraction experiments. These patterns were azimuthally integrated using pyFAI open-source software package [2]. X-ray diffraction peaks were fitted by Pseudo-Voigt function. The instrumental contribution to broadening of the peaks was calculated using diffraction pattern of LaB_6 standard and fitted by Caglioti function. Subsequent analysis of dislocation structure was performed by using deconvolution profile analysis approach described elsewhere [3].

3. Results and Discussion

Before performing plastic deformation, the sample was annealed and water-cooled. The scanning electron microscopy results shows the formation of a martensitic structure in alloy after heat treatment (Figure 1). Before plastic deformation the structure is represented by large grains and needle-shaped crystals.

Figure 1. The structure of Ti13Nb13Zr alloy before HPT. SEM.

According to the classical concepts presented by Il'in [4], the boundaries of the observed large grains correspond to the boundaries of the former grains of the β-phase of titanium. Besides, the plate-type (needle-shaped) crystals located in the grains are the martensitic α'-phase of the titanium. Thus, it can be considered that the structure of the material before plastic deformation is represented only by the titanium martensitic α'-phase.

The transmission electron microscopy results (Figure 2) show that during the plastic deformation some changes in phase composition occurs. The presence of α'- and β-phases was experimentally confirmed by using ACOM-TEM method (Figure 2 b). At the same time, results of other research groups [1] indicate the presence of a metastable hexagonal ω-phase (P6/mmm) in the Ti13Nb13Zr alloy after HPT deformation with 4.5 GPa loading.

The diffraction map of the deformed alloy is shown in the Figure 3. The obtained results confirm that in case of α'-titanium alloys the severe plastic deformation can lead to phase transitions. Firstly, the β-phase of titanium appears in the central area of the disk, which is consistent with the results of transmission electron microscopy. Secondly, obtained results indicate the presence of the ω-phase of titanium. However, this phase is present in small amounts (the ratio of intensities of the diffraction maxima (01.1)-ω to (10.2)-α' and (200)-β is 0.05 and 0.2, respectively).

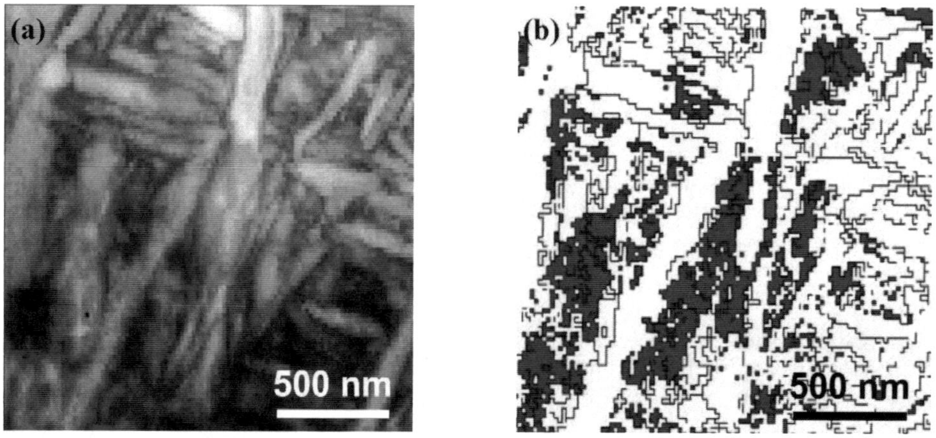

Figure 2. a – pseudo-light-field image; b – automatic phase distribution map constructed by the ACOM-TEM method of Ti13Nb13Zr alloy after HPT. The yellow color indicates the α'-phase, the blue color indicates the β-phase.

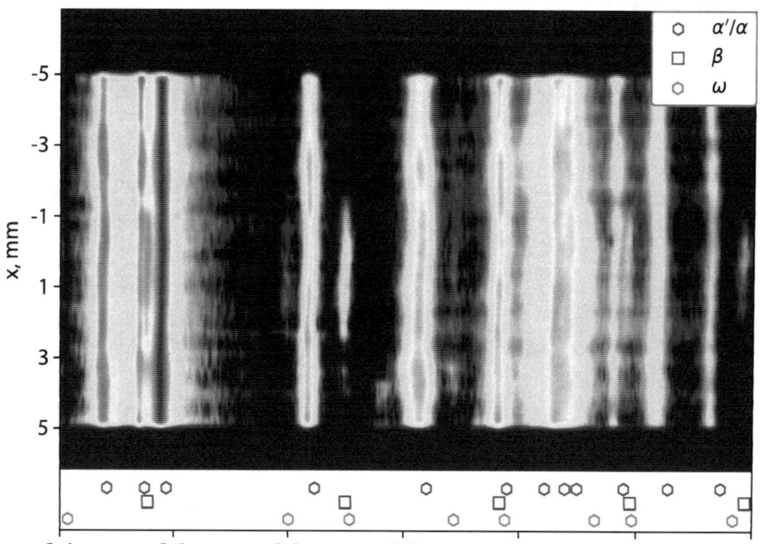

Figure 3. Diffraction map and phase composition of the Ti13Nb13Zr alloy after HPT processing along the sample diameter. Zero on the X axis corresponds to the center of the deformed disk.

Figure 4 shows the distribution of several structural parameters obtained by analyzing results of synchrotron X-ray diffraction data: size of coherent scattering regions ($<x_{area}>$); dislocation cut-off radius (R_e); Wilkens constant (M); total dislocation density (ρ) as well as density dislocations with of <a>, <c> and <c+a> Burgers vectors types; relative distribution of dislocations (h) with same Burgers vectors types. The results of the analysis of the dislocation structure allow to conclude that the dislocations distribution in the α'-phase is inhomogeneous. It should be noted that most common (from 40 to 60 %) type of dislocations is the medium-energy dislocations (<c> type of Burgers vectors). Simultaneously, the relative amount of low-energy dislocations <a> does not exceed 25 %.

The dislocation structure of the material is also heterogeneous along the deformed disk diameter. Low values of the Wilkens parameter indicate that the dislocations form ordered structures. In this case, the regions with the lowest dislocation density correspond to the maximum values of dislocations outer cut-off radius (R_e), which indicates a relatively low stress shielding of dislocations from each other.

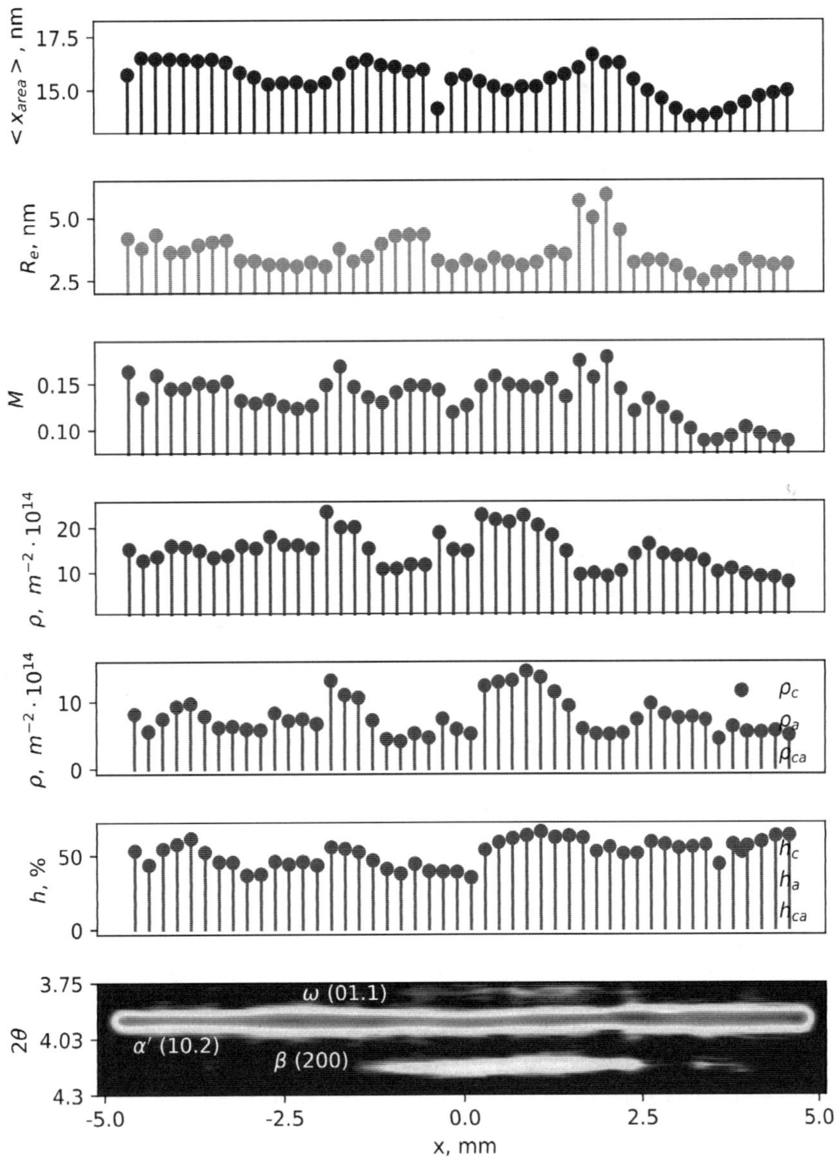

Figure 4. Distribution of structural parameters of the α'-phase of the Ti13Nb13Zr alloy along the diameter of the deformed disk relative to the diffraction maxima (01.1)-ω, (10.2)-α ' and (200)-β.

Figure 5 shows the result of the microhardness of the deformed disk surface. It should be noted that the uneven distribution of phases and structural parameters leads to a corresponding distribution of the microhardness of the material. The regions of the disk that are characterized by the presence of the β-phase, as well as the highest density of dislocations of the α'-phase, correspond to the highest values of microhardness.

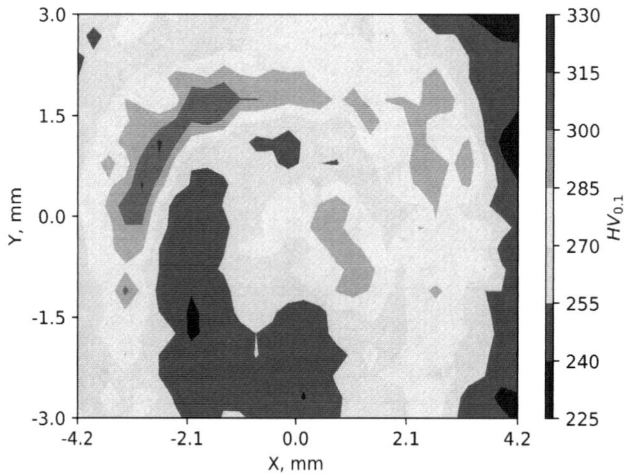

Figure 5. Microhardness distribution over the disk surface.

4. Conclusion

The approach used in the work based on scanning the diameter of the α'-titanium alloy disc showed that the high-pressure torsion leads to significant structure changes. It was found that the central zone of deformed material is characterized by the presence of β- and ω-phases. Furthermore, that local area possesses an increased dislocation density in the α'-phase compared with the peripheral zones of the disk. Finally, the distribution of phases and dislocation density in a severely deformed material has a significant effect on its microhardness.

Acknowledgments

This study was funded according to Russian Science Foundation research project № 20-73-10215 « In-situ study of the evolution of the dislocation structure of plastically deformed high-entropy alloys under high-pressures and temperatures using synchrotron radiation».

References

[1] Pérez D A G, Jorge Jr A M, Kiminami C S, Bolfarini C and Botta W J 2017 *Materials Research* **20** 404

[2] Ashiotis G, Deschildre A, Nawaz Z, Wright J P, Karkoulis D, Picca F E and Kieffer J 2015 *J. Appl. Crystallogr.* **48** 510

[3] Ivanov I V, Lazurenko D V, Stark A, Pyczak F, Thömmes A and Bataev I A 2020 *Metals and Materials Int.* **26** 83

[4] Il'in A A, Kolachev B A and Pol'kin I S 2009 *Titanium alloys. The composition, structure and properties* (Moscow: VILS-MATI) p 520

Formation of the CoCrCuFeNi high entropy cladding layer by non-vacuum electron beam treatment

A A Ruktuev[1], D V Lazurenko[1], N A Kiseleva[1], I Y Petrov[1] and M G Golkovski[2]

[1]Materials Science Department, Novosibirsk State Technical University, Karl Marks av. 20, Novosibirsk, 630073, Russia
[2]Budker Institute of Nuclear Physics of Siberian Branch Russian Academy of Sciences Lavrentiev av. 11, Novosibirsk, 630090, Russia

E-mail: ruktuev@corp.nstu.ru

Abstract. Non-vacuum electron beam cladding was applied to obtain CoCrCuFeNi cladding layers with thicknesses of 0.89 and 1.24 mm on a surface of mild steel. The cladding layers possessed the dendritic structure. Interdendritic space was filled with a Cu-rich phase. It was found by X-ray diffraction analysis that FCC solid solution phases were formed during crystallization. The lattice parameter of the dendritic phase was a=3.59 Å, while that of the interdendritic phase was equal to a=3.60 Å. The average microhardness values of the cladding layers were lower compared to the base material and equaled 156 HV and 190 HV.

1. Introduction

Recently, the multicomponent alloys with equiatomic and near equiatomic compositions, namely high entropy alloys, have attracted the attention of many research groups [1]. These alloys are a breakthrough in physical metallurgy since traditional alloys are usually based on one main component (steel, aluminum alloys, etc.).

Investigations of the high-entropy alloys have shown that this class of materials can have a unique combination of properties and find their application in industries that require high strength, hardness, wear resistance, or heat resistance. Due to these characteristics, high-entropy alloys can be considered as promising materials for protective functional coatings. A large number of studies are being carried out in this area, but most of them are aimed at studying thin films obtained by such technology as magnetron sputtering. However, it is more reasonable from the technological point of view to apply coatings of a large thickness (more than 0.5 mm). Various types of thermal spraying, laser cladding, arc cladding, and electron-beam cladding can be referred to as the methods of forming such thick coatings. Another currently important and actively developing technology is considered to be additive manufacturing with the use of the electron beam as an energy source [2]. Thus, the purpose of this study was to investigate regularities of the formation of materials with protective layers of the composition corresponding to high entropy alloys by the method of non-vacuum electron-beam processing. CoCrCuFeNi layers were applied to the surface of steel 20 workpieces. According to the literature data [3, 4], an alloy of such composition has an FCC structure and can be used at elevated temperatures, or as a corrosion resistant material.

2. Materials and methods

2.1. Preparation of the coatings

Formation of the surface layers was carried out at Budker Institute of Nuclear Physics using an ELV-6 electron accelerator modified to allow the electron beam to be injected into the atmosphere. The energy of electrons in the beam is 1.4 MeV. The advantages of this technology include the fact that energy is released in the subsurface layer. This makes it possible to efficiently melt a powder layer of considerable thickness applied to the surface.

To produce layers, a powder mixture consisting of commercially pure Co, Cr, Cu, Ni powders was used. The metal powders were mixed with a flux (CaF_2) powder in a ratio of 70:30 by weight (CoCrCuNi:CaF_2) and applied to the surfaces of steel 20 workpieces. Next, the powder-coated workpiece was moved forward under a scanning electron beam. Iron necessary for the formation of the cladding layer of the required elemental composition was received from the melt pool resulted from the melting of the base material. During the experiments, the amount of powder applied to the workpiece, the current of the electron beam, and the moving velocity of the workpiece under the beam were varied. Changing the modes allowed regulating the composition of the cladding layer and finally obtaining the equiatomic surface alloy. Table 1 shows the cladding regimes.

Table 1. Cladding regimes.

Regime No	Metal powders	Flux	Surface density, g/cm^2	Beam current, mA	Moving velocity, cm/s	Energy density J/cm^2
1	CoCrCuNi 70 %	CaF_2 30 %	0.45	25	15	4.67
2	CoCrCuNi 70 %	CaF_2 30 %	0.45	22	15	4.11
3	CoCrCuNi 70 %	CaF_2 30 %	0.45	20	15	3.73
4	CoCrCuNi 70 %	CaF_2 30 %	0.6	25	15	4.67
5	CoCrCuNi 70 %	CaF_2 30 %	0.8	25	15	4.67
6	CoCrCuNi 70 %	CaF_2 30 %	0.6	22	15	4.11
7	CoCrCuNi 70 %	CaF_2 30 %	0.8	22	15	4.11
8	CoCrCuNi 70 %	CaF_2 30 %	0.45	25	20	3.50

2.2. Characterization of the coatings

Cross-sections of the samples obtained were used for the microstructural investigations. Specimens were grinded and polished according to a standard method and etched with aqua regia (1HNO_3:3HCl). Microstructure was studied using a Carl Zeiss Axio Observer Z1m optical microscope (OM), and a Carl Zeiss EVO50 scanning electron microscope (SEM). Elemental compositions were measured by energy-dispersive X-ray spectroscopy using an Oxford Instruments X-Act spectrometer coupled with SEM. Phase compositions and lattice parameters were estimated using a Arl X'tra X-ray diffractometer. X-ray diffraction (XRD) patterns were obtained in Cu Kα1+2 irradiation. 2Θ ranged between 30 and 155 °, and a step size was equal to 0.03 °. Samples for the XRD analysis were prepared by grinding and polishing the top surface of cladding layers to the depth ~0.3 mm. Profile analysis was implemented using WinXRD software. Peak profiles were described by Pearson VII function in consideration of Cu Kα2 irradiation. Lattice parameters were calculated using the Nelson-Riley function [5].

The Vickers microhardness was measured on the polished cross-sectional specimens in the direction from the top of the cladding layer down to the substrate by a Wolpert Group 402MVD hardness tester using a 100 g load on a diamond indenter.

3. Results and discussion

Energy dispersive analysis of the cladding layers revealed that modes 3 and 5 provided the composition closest to the required equiatomic one. 1st, 2nd, 4th, and 6th regimes caused excessive melting of the base material, which led to the formation of layers with an iron content of 30 - 75 at. %. Contrary, regimes 7 and 8 did not provide the required depth of melting of the base material. The elemental composition and the thickness of the cladding layers obtained according to the 3rd and 5th regimes are listed in Table 2. The copper concentration was lower than expected, which can be explained by insufficient protection of the molten bath from air oxygen and more intense oxidation of copper compared to other elements.

Table 2. Average elemental compositions of the cladding layers.

№	Co, at. %	Cr, at. %	Cu, at. %	Fe, at. %	Ni, at. %	Thickness, mm
3	20	21	18	21	20	0.89
5	21	22	17	20	20	1.24

Figure 1 shows a typical microstructure of the materials' cross-sections. Three main regions can be distinguished in the structure: a cladding layer, a heat-affected zone, and a base material.

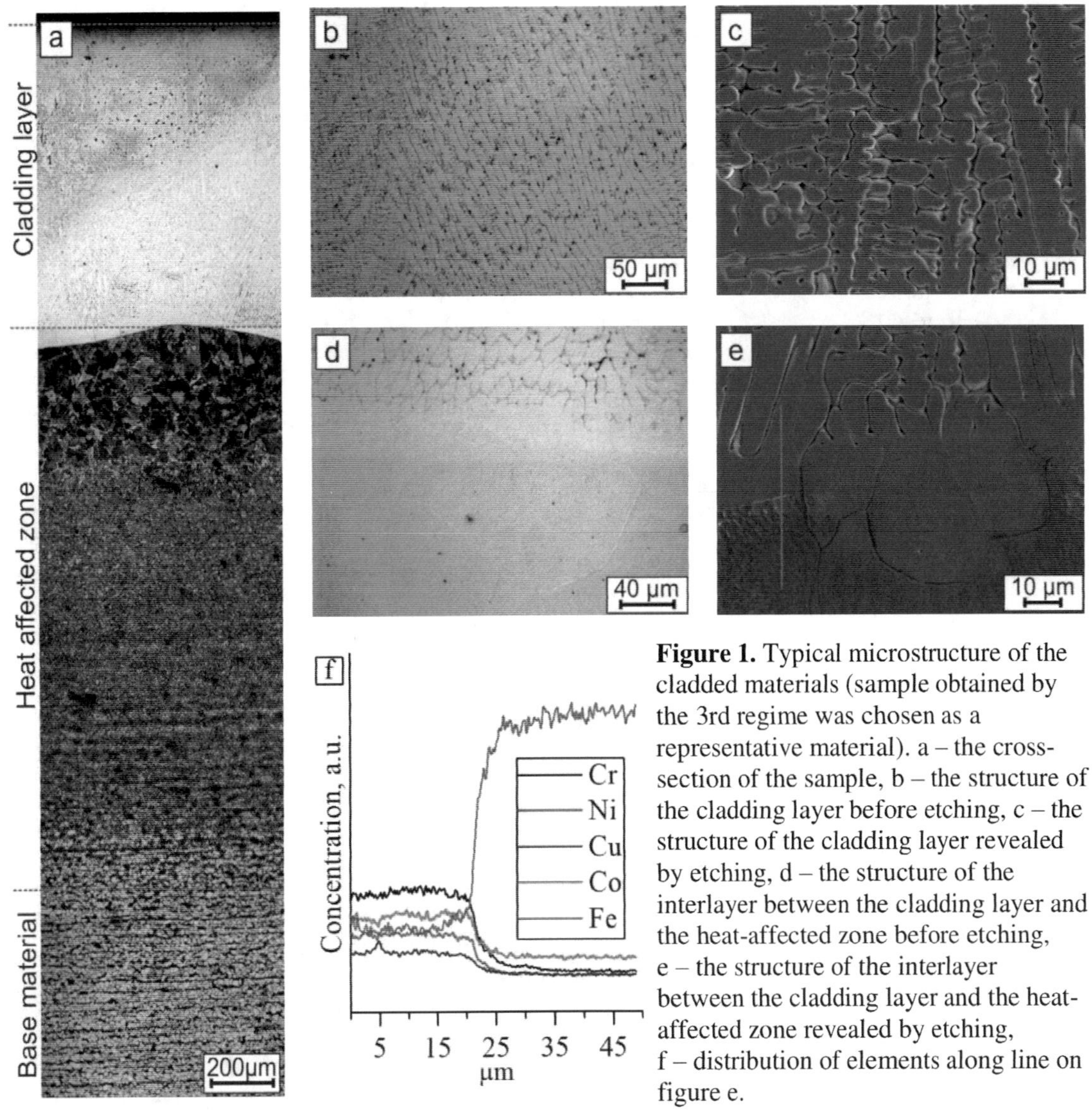

Figure 1. Typical microstructure of the cladded materials (sample obtained by the 3rd regime was chosen as a representative material). a – the cross-section of the sample, b – the structure of the cladding layer before etching, c – the structure of the cladding layer revealed by etching, d – the structure of the interlayer between the cladding layer and the heat-affected zone before etching, e – the structure of the interlayer between the cladding layer and the heat-affected zone revealed by etching, f – distribution of elements along line on figure e.

The cladding layer had a dendritic structure varying in depth. The lower part of the layer corresponded to the cellular growth of crystallites (Figure 1 d,e). Due to intensive heat removal into the base material, a large number of crystallization nuclei was formed at the border with the unalloyed metal. At some distance from this boundary, the structure was typical dendritic (Figure. 1b and 1c): the dendrites in the central part were elongated in the vertical direction, which also can be explained by more intense heat removal towards the base material. It should be noted that a thin interlayer with a thickness of ~10 μm was formed between the cladding layer and a heat-affected zone. In this area, a gradual decrease in the concentration of alloying elements and an increase in iron concentration was observed (Figure 1f). Another peculiarity of the structure of the interface consisted in the formation of local areas where «leakage» of a Cu-enriched phase in the heat-affected zone was observed (Figure 1 d, e). In such areas, the crystallite boundaries extending from the cladding layer to the heat-affected zone can be clearly distinguished.

A similar crystallization character was observed in [6]. In addition, in the cladding layers obtained in our study, as well as in the aforementioned work, the vertical cracks extending along the dendrite boundaries were formed. The formation of these cracks can be associated with the relaxation of stresses arising in the base material during heating in a process of the electron beam cladding.

Energy dispersive analysis revealed that dendrites formed in the cladding layer were depleted in copper compared to the average composition, while the Cu-rich phase crystallized in the interdendritic space (Table 3). The copper concentration reached 65 – 75 at. % in the interdendritic space.

Table 3. Local elemental compositions and lattice parameters.

Sample	Region	Co, at. %	Cr, at. %	Cu, at. %	Fe, at. %	Ni, at. %	a, Å
3	Dendrite	23.0	22.4	10.4	24.2	19.9	3.59
	Interdendrite	5.5	6.4	71.1	6.4	10.9	3.60
5	Dendrite	24.7	24.7	12.4	15.9	22.2	3.59
	Interdendrite	7.2	7.9	65.4	5.4	14.0	3.60

Figure 2 shows XRD patterns obtained from the cladding layers. It was found that the materials had an FCC structure.

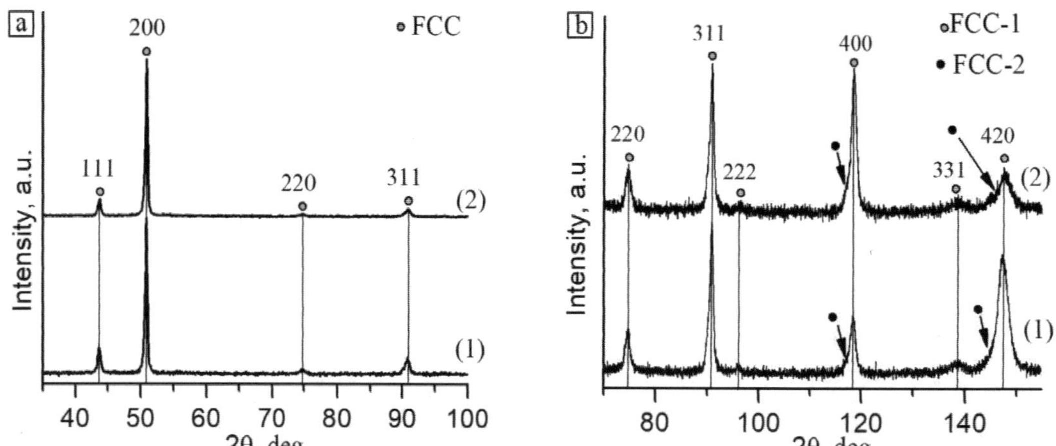

Figure 2. XRD patterns obtained from: 1- coating No 3, 2 - coating No 5. a - 2Θ = 30 – 100 degrees, b - 2Θ = 70 – 155 degrees measured for lattice constant calculations.

Detailed analysis revealed a slight asymmetry of the peaks clearly visible at the reflections corresponding to the families of planes (400) and (420) (Figure 2b). It was assumed that the intense peaks corresponded to the dendritic phase (FCC-1), while the asymmetry is due to the presence of an interdendritic Cu-rich solid solution (FCC-2). Based on these assumptions, a profile analysis was carried out and the lattice parameters of the dendritic and interdendritic phases were calculated (Table 3).

Microhardness of the cladding layers obtained by regimes 3 and 5 was 156±3 HV and 190±3 HV respectively. These values were lower than that of the base material (220±4 HV). The absence of an increase in hardness according to the solid solution hardening model can be explained by the close values of the atomic radii of the elements used for alloying.

Conclusions

Non-vacuum electron beam cladding allows obtaining the CoCrCuFeNi high-entropy surface layers at carbon steel workpieces. The thickness of layers obtained in one pass can reach 1.24 mm.

The cladding layer has a dendritic structure. The interdendritic space consisted of a copper-based phase with the concentration of other elements at a level of 5-10 % at.

Dendrites and the interdendritic space were represented by FCC solid solutions with similar lattice parameters.

Acknowledgments

Materials characterization was carried out at NSTU core facility center: “Structure, mechanical and physical properties of materials”

This research was carried out using financial support from the Council on grants of the President of the Russian Federation for young scientists (grant number MK-2226.2020.8)

References

[1] George E P, Raabe D and Ritchie R O 2019 High-entropy alloys *Nature Reviews Materials* **4** 515-34

[2] Tarasov S Y, Filippov A V, Shamarin N N, Fortuna S V, Maier G G and Kolubaev E A 2019 Microstructural evolution and chemical corrosion of electron beam wire-feed additively manufactured AISI 304 stainless steel *J. Alloys Compd.* **803** 364-70

[3] Cheng J B, Liang X B, Wang Z H and Xu B S 2013 Formation and Mechanical Properties of CoNiCuFeCr High-Entropy Alloys Coatings Prepared by Plasma Transferred Arc Cladding Process *Plasma Chem. Plasma Process.* **33** 979-92

[4] Zhang H, He Y Z, Pan Y and Guo S 2014 Thermally stable laser cladded CoCrCuFeNi high-entropy alloy coating with low stacking fault energy *J. Alloys Compd.* **600** 210-4

[5] Nelson J B and Riley D 1945 An experimental investigation of extrapolation methods in the derivation of accurate unit-cell dimensions of crystals *Proceedings of the Physical Society* **57** 160

[6] Lazurenko D V, Alferova G I, Golkovsky M G, Emurlaev K I, Emurlaeva Y Y, Bataev I A, Ogneva T S, Ruktuev A A, Stepanova N V and Bataev A A 2020 Formation of wear-resistant copper-bearing layers on the surfaces of steel substrates by non-vacuum electron beam acladding using powder mixtures *Surf Coat Tech* **395**

Deformation of silumin subjected to a pulsed electron beam treatment

M E Rygina[1,2], A M Ustinov[3,4], E A Petrikova[1], A A Klopotov [3], Yu F Ivanov[1], A D Teresov[1], Yu A Abzaev[3], L M Loskutov[3] and G G Volokitin[3]

[1]Institute of High Current Electronics of the Siberian Branch of the Russian Academy of Sciences (SB RAS), 634055, Tomsk, Russia
[2]Tomsk Polytechnic University, 30 Lenin Ave., 634050, Tomsk, Russia
[3]Tomsk State University of Architecture and Building, sq. Solyanaya 2, 634003, Tomsk, Russia
[4]Siberian State Industrial University, 42 Kirova str., 654007, Novokuznetsk, Russia

E-mail: yufi55@mail.ru

Abstract. Using a virtual extensometer, it was found that irradiation with a pulsed electron beam (18 keV, 15 J/cm^2, 200 μs, 3 pulses, 0.3 s^{-1}) leads to an increase in the strength properties of the hypoeutectic silumin samples by 30%. The deformation curves in the coordinates «σ -ε» and «σ - √ε» did not reveal inflection points that characterize the stages of deformation transformation of the material. This indicates that the transitions from stage to stage in this class of materials are blurred.

1. Introduction

Almost all wear, corrosion, growth of fatigue cracks processes etc., leading to the failure of parts and mechanisms, begin from the surface and are determined by the properties of a relatively thin surface layer [1]. Consequently, in some cases, to increase the product service life, it is sufficient to modify its surface layer. Currently, methods based on the use of concentrated energy flows, including combined electron-ion-plasma methods are promising, demonstrating the high efficiency of products surface modification [2].

The aim of this work is to establish and analyze the regularities of plastic deformation of silumin irradiated by an intense pulsed electron beam and subjected to tensile deformation.

2. Materials and method

Silumin of AK10M2N brand, widely used in mechanical engineering, aerospace industry, shipbuilding, instrument making, medicine, etc., was used as an object of modification [3]. Modification of silumin specimens was carried out on a SOLO setup [4]. Mechanical tests of silumin specimens were carried out by stretching proportional flat specimens with heads in accordance with the International Standard "ISO 6892-84. Metals. Methods of tension test", that regulates the dimensions of the specimens used for tensile tests [5]. The electron beam was oriented along the normal to the specimen plane during irradiation. The surface of the samples was irradiated from both sides with an electron beam with the parameters 18 keV, 15 J/cm^2, 200 μs, 3 pulses, 0.3 s^{-1}.

The study of the processes occured during silumin deformation was carried out using the contactless method of digital image correlation (DIC) [6]. Using this method, with a virtual extensometer, the deformation curves were obtained by uniaxial tension of flat specimens. The use of the digital image correlation method made it possible to analyze the images of the investigated surface of a solid specimen to obtain qualitative and quantitative characteristics of the deformations fields, displacements, and velocities. The deformation pattern of a surface is obtained by combining changes in sub-regions. Speckle structures are artificially applied to the surface of solid specimen to form a contrasting pattern. The Vic3D digital optical system uses the principle of digital stereoscopic images correlation, that makes it possible to obtain data on the movement of the sample surface in space. Shooting from two digital cameras allows obtaining a stereoscopic image of the specimen surface. Analyzing them, the movement of points on the specimen surface in three axes (in space) is determined. Using the digital image correlation method implemented in the Vic3D digital optical system, the geometric parameters of the surface (X, Y, Z coordinates for each analyzed point) are calculated, as well as displacement at each point (U, V and W, indicating displacements along the X axes, Y and Z, respectively), relative deformations ((ε_{xx} - along the X axis, ε_{yy} - along the Y axis, ε_{xy} - shear deformations), the rate of change in displacement and deformations, surface curvature. paragraphs are indented.

3. Results and discussion

It was found that irradiation of silumin with a pulsed electron beam in the melting mode of a relatively thin (up to 20 μm) surface layer and subsequent high-speed crystallization lead to the formation of a cellular crystallization structure (solid solution based on an aluminum crystal lattice) with dimensions (200-250) nm. Nanosized (30-50 nm) particles of silicon and intermetallic compounds are located in the volume and along the boundaries of the cells [7].

The figure 1 shows the deformation curves in coordinates σ from ε and σ from √ε for specimens before irradiation and after irradiation with an intense pulsed electron beam. It is seen that irradiation of silumin with a pulsed electron beam makes it possible to increase the plasticity of the specimen.

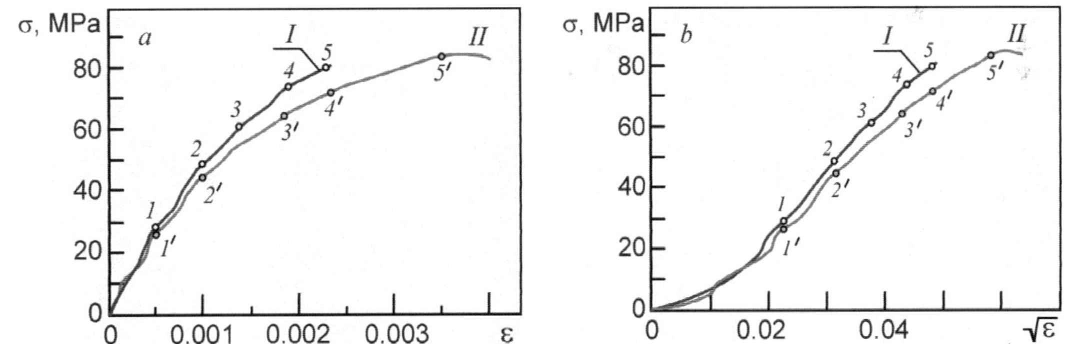

Figure 1. Tensile strain curves of silumin specimens in σ coordinates from ε (*a*) and from √ε (*b*) before irradiation (I) and after irradiation with an intense pulsed electron beam (II).

The staging of deformation curves by processing the hardening curves in coordinates σ–ε^n is the standard approach. The inflection points on the deformation curves can be used to judge the structural changes occurring in the alloys. The presence of inflection points on the deformation curves in the σ–√ε coordinates indicates structural changes during deformation. There are no characteristic inflection points on the deformation curves in the coordinates «σ – ε» and «σ – √ε» (figure 1). in the studied silumin. Their absence indicates that the transitions from stage to stage in this class of materials are blurred.

On the other hand, the non-monotonic nature of the change in the given deformation curves in the coordinates «σ – ε» and «σ – √ε» reflects the temporary inhomogeneity of the process of plastic deformation of the alloy. This phenomenon is accompanied by localization of deformation processes at various scale and structural levels [8] and is confirmed by the deformation field distributions patterns on the surface of the samples under study depending on the deformation degree (figures 2-3). The figure 2 shows the pictures of the deformation fields distributions for points 1-5 indicated in the diagrams in the figure 1. Macrovolumes located in a chaotic manner are observed, in which plastic deformation of the material occurs. At the initial stage, the centers of local deformation are formed mainly on the lateral sides of the sample. With an increase in the applied stress, they come into contact with the formation of the so-called deformation channels located at an angle of 45 degrees to the lateral surface of the specimen [9].

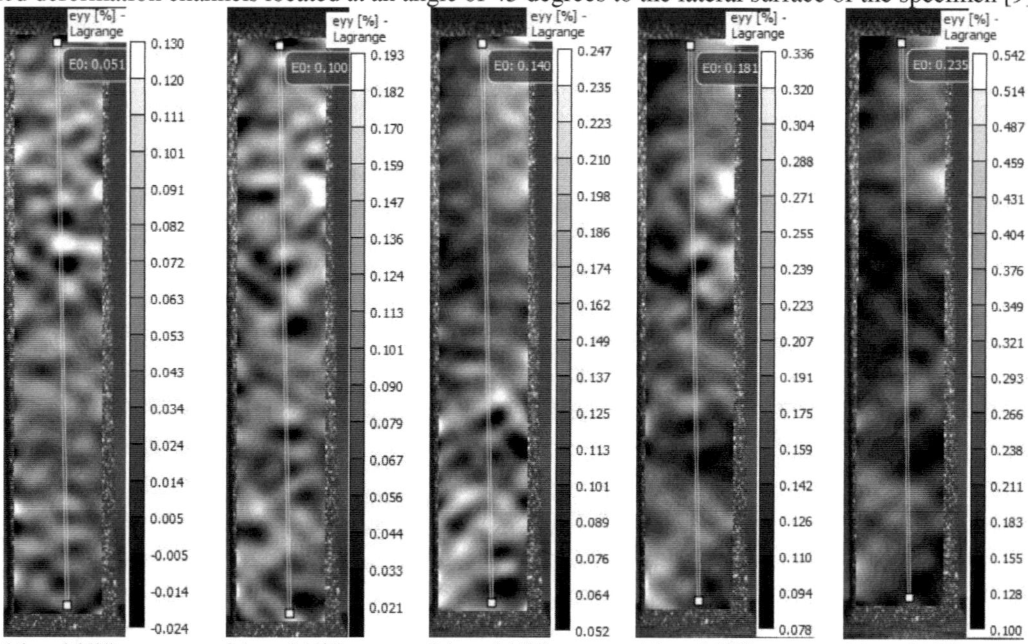

1 (ε_{YY}=0.05%) 2 (ε_{YY} =0.10%) 3 (ε_{YY} =0.14%) 4 (ε_{YY} =0.18%) 5 (ε_{YY} =0.23%)

Figure 2. Pictures of longitudinal distributions ε_{YY} relative deformations on the surface of an unirradiated silumin specimen. Patterns 1 - 5 correspond to points 1 - 5 on the deformation curve I in figure 1.

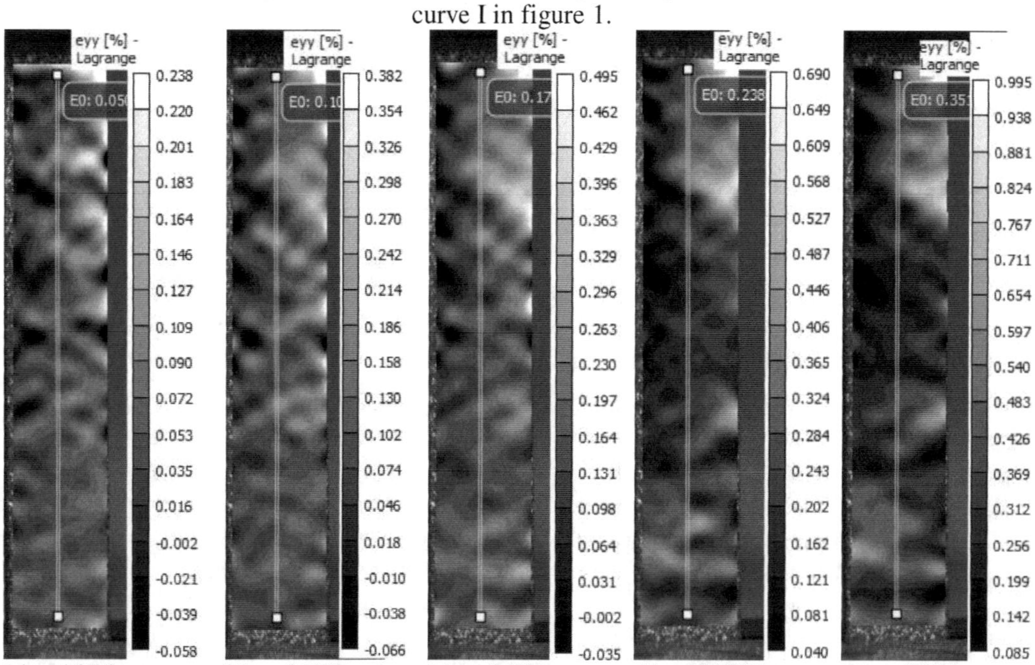

1 (ε_{YY}=0.05%) 2 (ε_{YY} =0.10%) 3 (ε_{YY} =0.17%) 4 (ε_{YY} =0.238%) 5 (ε_{YY} =0.35%)

Figure 3. Pictures of longitudinal distributions ε_{YY} relative deformations on the silumin specimen surface irradiated with a pulsed electron beam. Patterns 1 –5 correspond to points 1′–5′ on a deformation curve II in figure 1.

The deformation fields distribution patterns during stretching of the specimen irradiated with an electron beam (figure 3) are similar to the deformation fields distributions patterns on the original sample (figure 2). As in the case of deformation of unirradiated silumin, stretching of the irradiated specimen is accompanied by the formation of regions with high values of longitudinal deformations ε_{YY} at the last stage of the deformation curve, localized on the lateral face of the specimen. The difference is the formation of compression zones on the surface between tensile local deformation zones (figure 3).

Further study of the features of the effect of irradiation on the mechanical properties of silumin specimens was carried out by plotting diagrams in coordinates $\varepsilon_{YY}^{max} - \sigma$ and $\varepsilon_{YY}^{min} - \sigma$. Here ε_{YY}^{max} the maximum values of deformations inside the centers of plastic deformation in the longitudinal direction; ε_{YY}^{min} minimum values of deformations in the longitudinal direction in the accommodation zones (figure 4).

It can be seen that with an increase in the external applied tensile stress ε_{YY}^{max}, they increase in a non-linear manner in the initial and irradiated samples (figure 4, a). In irradiated samples, the increase ε_{YY}^{max} occurs more intensively with an increase in the applied tensile stress. In zones with minimal deformations values in the longitudinal direction, an oscillating character of the change in deformations from the applied stress is observed with a general tendency towards an increase in numerical values (figure 4, b). This indicates that accommodative processes occur in the main zones with minimal deformation values.

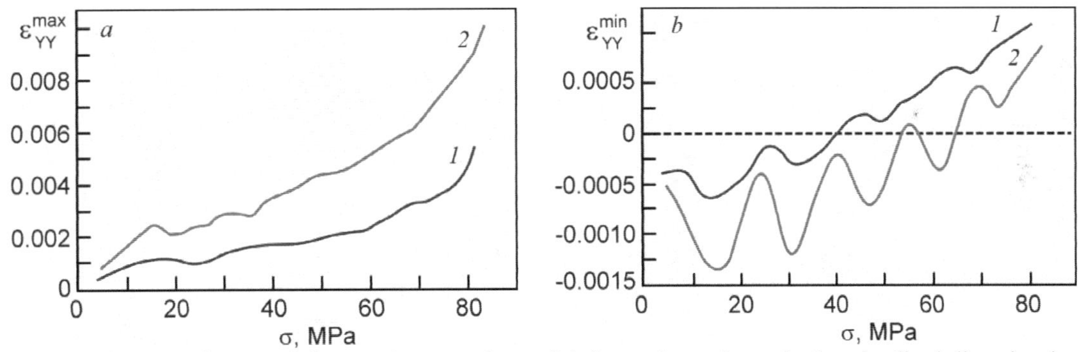

Figure 4. Dependences of the maximum values of deformations along the longitudinal direction in the area of the centers of plastic deformation ε_{YY}^{max} (a) and in the area of minimum values of deformations ε_{YY}^{min} (b). 1 – original sample; 2 – sample irradiated with a pulsed electron beam.

4. Conclusion

Using a virtual extensometer, it was found that irradiation with a pulsed electron beam (18 keV, 15 J/cm^2, 200 μs, 3 pulses, 0.3 s^{-1}) leads to an increase in the strength properties of the hypoeutectic silumin samples by 30%. The deformation curves in the coordinates «σ – ε» and «σ – √ε» did not reveal inflection points that characterize the stages of deformation transformation of the material. This indicates that the transitions from stage to stage in this class of materials are blurred. The non-monotonic nature of the change in the deformation curves in the coordinates «σ – ε» and «σ – √ε» is established, that reflects the temporal inhomogeneity of the process of inelastic deformation of the alloy. The manifestation of this process correlates well with the evolution of deformation fields on the surface of the samples. Moreover, this phenomenon is most clearly observed on irradiated samples.

Acknowledgments
This work was supported by the RFBR-BRFBR grant (project 19-52-04009) (irradiation of silumin with an electron beam and study of its structure) and by the Russian Science Foundation

(project 19-79-10059) (study of the processes occurring during the deformation of silumin carried out using non-contact digital image correlation (DIC)).

References

[1] Ivanov Yu F, Gromov V E, Konovalov S V and Aksenova K V 2019 *Fatigue of Electron Beam Modified Silumin* (Novokuznetsk: Publishing house "Polygraphist") p 184

[2] Laskovnev A P, Ivanov Yu F, Petrikova E A and other 2013 *Modification of the structure and properties of the eutectic effect by electron-ion-plasma treatment* (Minsk: Belarus. Navuka) p. 287

[3] Zolotarevsky V S and Belov N A 2005 *Physical metallurgy of cast aluminium alloys* (Moscow: MISiS) p 376

[4] Koval N N, Grigoriev S V, Devyatkov V N, Teresov A D and Schanin P M 2009 Effects of intensified emission during the generation of a submillisecond low-energy electron beam in a plasma-cathode diode *IEEE Trans. on Plasma Sci.* 37(10) pp 1890-1896.

[5] GOST R 1497-84 2005 *Metals. Tensile test methods* (Moscow: Standard Publishing House)

[6] Tretyakova T V and Wildemann V E 2018 Plastic Strain Localization and its Stages in Al-Mg Alloys *Phys. Mesomech.* **21(4)** pp 314-319.

[7] Ivanov Yu F, Zagulyaev D V, Nevskii S A, Gromov V E, Sarychev V D and Semin A P Microstructure and Properties of Hypoeutectic Silumin Treated by High-Current Pulsed Electron Beams *Prog. in Phys. of Metals* **20** pp 451–490.

[8] Panin V E and Egorushkin V E 2008 Nonequilibrium thermodynamics of a deformed solid as a multilevel system. Corpuscular Wave Dualism of Plastic Shear *Phys. Mesomech.* **11(2)** pp 9-30

[9] Farber V M, Polukhina O N, Vichuzhanin D I, Khotinov V A, Smirnov S V 2019 A study of plastic deformation of 08G2B steel before and at the yield plateau by digital image correlation technique. Part 1. Formation of plastic and elastic deformation waves *Melal Science and Heat Treatment* **61** pp 274-280

Research of the damage accumulation and destruction processes of perforated composite plates

E M Strungar, E V Feklistova, V E Wildemann and D S Lobanov

Perm National Research Polytechnic University, Perm

E-mail: cem.spaskova@mail.ru

Abstract. The main purpose of the research is to comprehensively study the regularities of deformation processes damage accumulation and destruction of modern composite materials in stress concentration zones. The object of research is fiberglass samples with four round holes. As a result of experimental studies loading diagrams were obtained for samples with different length of the working area. The paper presents graphs of the load dependence on the longitudinal deformation. According to the data obtained it can be noted that the longer the length of the working area of the sample the earlier the failure occurs. The deformation was obtained using the optional Vic-3D "virtual extensometer" non-contact optical video system. In the experiments process curves of longitudinal deformations were plotted. From these data it is possible to estimate the maximum longitudinal deformations occurring near the hole. Also in the work numerical modeling of the process of deformation and destruction of perforated plates was carried out in the software engineering complex Ansys. The calculated data obtained confirm the experimental results of the research.

1. Introduction

In industry, in particular in the civil aviation industry the connection of composite elements in the structure is carried out using mechanical fastening. This process requires a large number of drilled holes for location the rivets and bolts. Assembly work is carried out using fasteners. The staggered arrangement of holes for bolted joints is often used both in metal structures and in composite ones, since this allows to increase the strength of structures [1].

With the use of the Vic-3D non-contact optical video system, the mathematical apparatus of which is based on the DIC digital image correlation method, it becomes possible to study the stress-strain state of the material in the area of the concentrator, at the point of crack initiation, and also to characterize the state that precedes fracture [2-5]. Measurements of the total strain field using the digital image correlation method are very important for checking the strain distribution in a loaded specimen where there is a local contact between the bolt and the hole [6, 7]. The processes of structural destruction are reflected in the deformation diagram leading to its nonlinearity, and at the final stage are the cause of softening [8-13]. The fracture resistance at the post critical stage of deformation corresponding to the descending branch of the deformation diagram depends on the rigidity of the loading system. Numerical modeling of the fracture process is performed using various fracture criteria and degradation mechanisms of material properties. The works [14-18] use the fracture criterion based on stress and strain, in [19-20] on the basis of energy.

2. Material, sample and test equipment

This work was carried out in Perm National Research Polytechnic University using Unique Scientific Equipment «Complex of testing and diagnostic equipment to study the properties of structural and functional materials under complex thermomechanical loading conditions» http://ckp-rf.ru/usu/501309/" at the Center experimental mechanics.

The main purpose of the research is to comprehensively study the regularities of deformation processes damage accumulation and destruction of modern composite materials in stress concentration zones. The experiments were carried out on an Instron 5989 universal electromechanical testing system. Registration of inhomogeneous deformation fields on the sample surface was carried out using a Vic-3D digital optical system based on the DIC digital image correlation method. The experiments were carried out on an Instron 5989 universal electromechanical testing system. Registration of inhomogeneous deformation fields on the sample surface was carried out using a Vic-3D digital optical system based on the DIC digital image correlation method. The speed of the moving gripper was u'_0=2 mm/min. The load was measured with a load cell with an ultimate load of 100 kN. The load measurement accuracy is not less than 0.5% of the measured value. The video system was implemented using two digital monochrome cameras Prosilica GE4900 (lens - 35 mm f / 2.0), the shooting speed was 4 frames per second.

The research program included uniaxial tensile mechanical tests of composite specimens 100 mm wide and 4 mm thick with 4 round holes of equal diameter d = 6 mm and with different length of the working area l_{work} = 150, 200, 240 mm. Different sample lengths were chosen to study the effect of the rigidity of the loading system. A total of two samples of each length were tested. Fiberglass plates were made of GFRP (1.95 mm thick) manufactured in accordance with GOST (National Standard) 12652-74. This material consists of a plain weave fabric (6 layers) and a hot cure phenolic epoxy matrix.

3. Main part

3.1 Results of the experimental study

As a result of experiments, loading diagrams were obtained for perforated specimens with different length of the working part (figure 1). From the data obtained, it can be seen that on the curves for specimens with a minimum working base of 150 mm on the falling branch, in addition to the dynamic section of destruction, there is an equilibrium section. As the length of the working section increases to 200 mm and 240 mm, a gradual dynamic destruction of the sample occurs, while the equilibrium section is practically absent.

Figure 2 shows a graph of the load on the longitudinal deformation dependence, the curves shown are presented with a displacement of 0.02 mm. It should be noted that the deformation was obtained with the help of an additional instrument of the video system «virtual extensometer». With the help of it, it becomes possible to obtain deformations without mechanical impact on the surface of the sample and to apply it in any area of interest. According to the data obtained, it is possible to note the longer the length of the working part of the sample, the earlier the break of the diagram occurs.

During the experiments, plots of longitudinal deformations were obtained along the line get through the holes in the horizontal direction in the images - figure 3 over the entire width of the sample 100 mm at the moment corresponding to the maximum load level. In this case, the Oy axis is directed along the sample (along the tension axis), and the Ox axis is perpendicular to the loading axis in the sample plane.

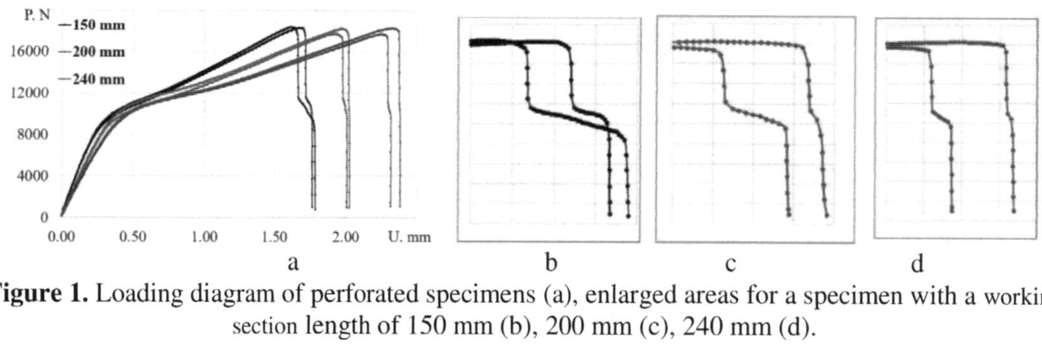

Figure 1. Loading diagram of perforated specimens (a), enlarged areas for a specimen with a working section length of 150 mm (b), 200 mm (c), 240 mm (d).

IOP Conf. Series: Materials Science and Engineering 1093 (2021) 012030 doi:10.1088/1757-899X/1093/1/012030

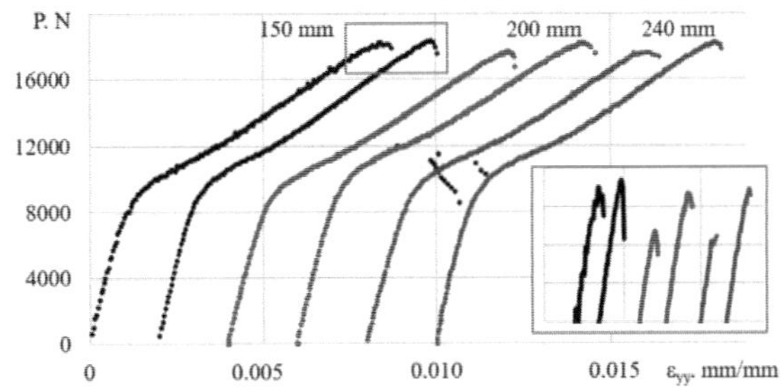

Figure 2. Load versus longitudinal strain diagram for perforated specimens.

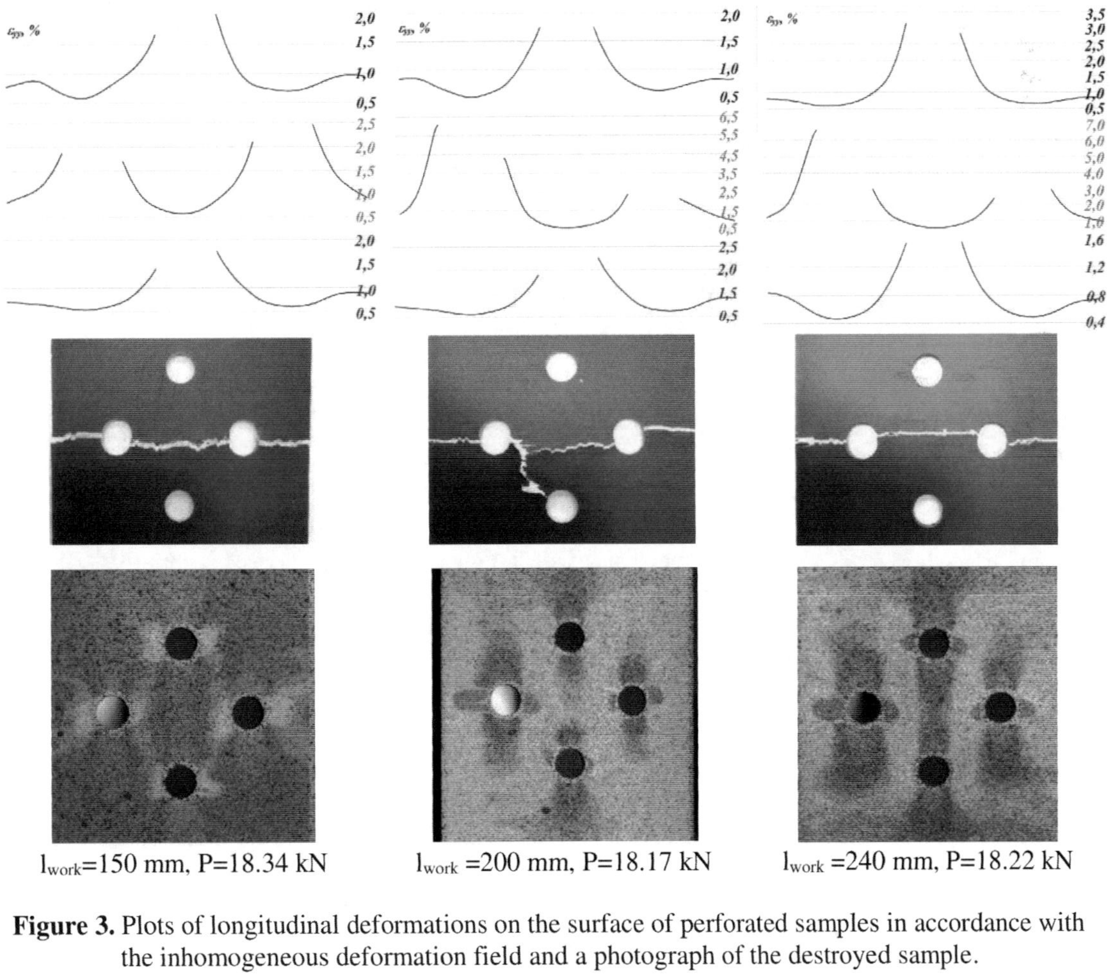

l_{work}=150 mm, P=18.34 kN l_{work} =200 mm, P=18.17 kN l_{work} =240 mm, P=18.22 kN

Figure 3. Plots of longitudinal deformations on the surface of perforated samples in accordance with the inhomogeneous deformation field and a photograph of the destroyed sample.

The presence of the hole leads to a rupture of 6 mm wide, which corresponds to their diameter. From the data presented, it is possible to estimate the maximum longitudinal deformations occurring near the hole. The given field of longitudinal deformations at maximum load clearly demonstrates the location of defects.

According to all the results obtained, it can be noted that in 5 cases out of 6 the destruction of the sample occurred as a result of the development of a zigzag crack along the smallest transverse area.

Taking into account this geometry of the staggered arrangement of the holes, it can be concluded that the two additional holes did not affect the destruction process. As an example, figures 4 show inhomogeneous fields of transverse, shear and longitudinal deformations on the surface of a perforated sample with a working section length of 240 mm at different times. From the presented inhomogeneous fields, it is possible to establish the presence and increase in the concentration of the level of deformations around the hole, which in turn leads to the growth of a crack and subsequent destruction of the sample.

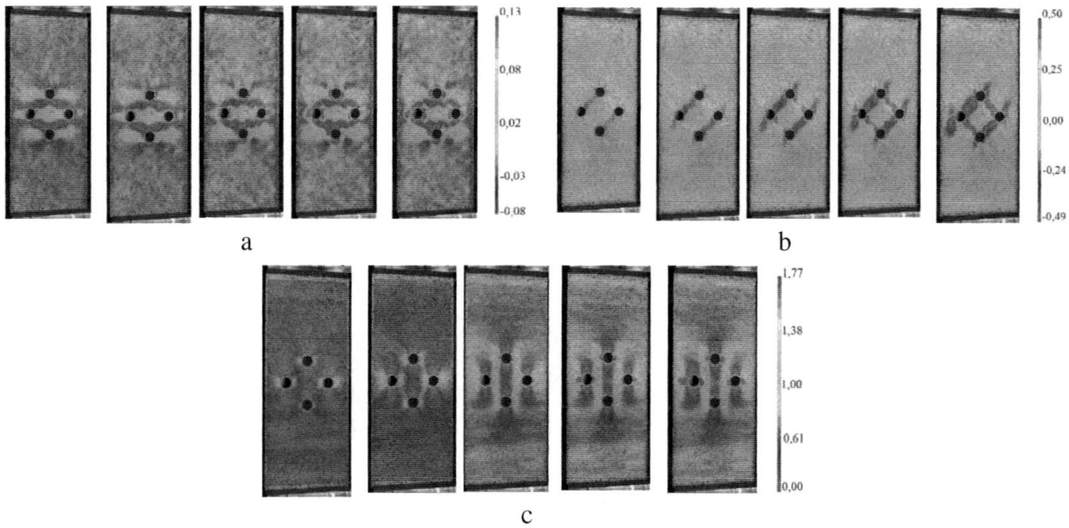

Figure 4. Transverse strain fields (a), fields of shear (b) and longitudinal (c) deformations for a specimen with a working section length of 240 mm.

3.2 Results of numerical calculations

Numerical modeling of deformation and destruction process of a perforated plate with a width of H = 100mm, a length of L = 350mm and four holes with a radius of R = 6mm has been carried out. The mathematical model of the deformation and fracture process is described by a system of equations (1)-(3):

$$\sigma_{ij,j}(r)=0 \quad (1)$$

$$\varepsilon_{ij}(\mathrm{r}) = (u_{i,j}(r)+u_{j,i}(r))/2 \quad (2)$$

$$\sigma_{ij}(r)=[3K(r)(1-k)\,V_{ijmn}+2G(1-g)\,H_{ijmn}]\,\varepsilon_{ij}(r) \quad (3)$$

in which the functions g and k describe elastic-brittle fracture varying from 0 to 1 in an abrupt manner when the fracture criterion is met:

$$g\left(j_\sigma^{(2)}\right)=\begin{cases}0, j_\sigma^{(2)} < j_{\sigma\,cr}^{(2)}\\ 1, j_\sigma^{(2)} > j_{\sigma\,cr}^{(2)}\end{cases}, \quad k\left(j_\sigma^{(1)}\right)=\begin{cases}0, j_\sigma^{(2)} < j_{\sigma\,cr}^{(2)}\\ 1, j_\sigma^{(2)} > j_{\sigma\,cr}^{(2)}\end{cases}$$

The system of equations is supplemented by kinematic boundary conditions on the body surface S: $U_i(r)|s = U_i^0$ [10].

In this case, the fracture process is modeled by comparing the largest principal stresses σ_{11} with the ultimate stress σ_{cr}, which is the same for all finite elements and with an automatic selection of the loading step. The numerical model is built in the software engineering complex Ansys. The solution of the problem is carried out in a plane-stressed formulation using a four-node finite element Plane 182. The material of the plate is taken to be orthotropic, the properties of which are given in Table 1 [21].

Loading was carried out by applying a displacement along the vertical axis to the upper face of the plate, the lower face was fixed along the loading axis, and the lower left point was fixed rigidly.

Table 1. Material properties of the perforated plate.

E_{xx}, Pa	E_{yy}, Pa	υ_{xy}	G_{xy}, Pa
$28 \cdot 10^9$	$28 \cdot 10^9$	0.18	$6.2 \cdot 10^9$

As a result of the solution, the calculated pictures of the fracture zones were obtained (figure 7), built for various moments of deformation of the plate with concentrators.

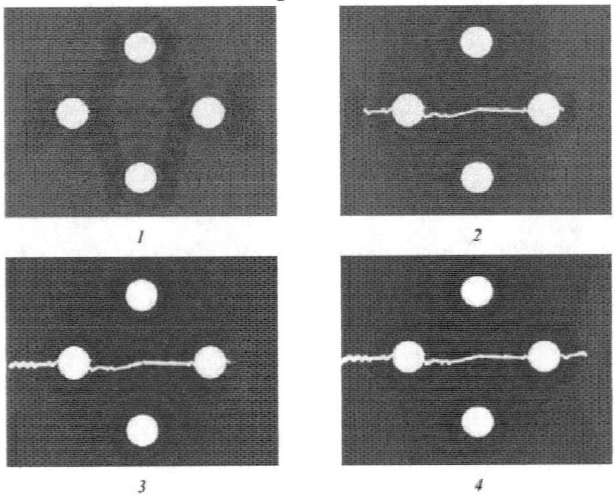

Figure 5. Pictures of fracture zones plotted for various moments of deformation of a plate with concentrators.

The crack in all specimens propagates perpendicular to the load application axis. The calculated strain distribution fields, as well as the direction of crack propagation, are in qualitative agreement with the experimental results. The photographs of the type of failure shown in figure 5 are consistent with the calculated and experimental fields, and correspond to the failure of the fibers in the transverse direction in the area of the stress concentrator.

4. Conclusion

Based on the results of experimental studies of perforated fiberglass plates with round concentrators, the effect of the loading system on the plate destruction process was estimated. As the length of the test section increased, a gradual dynamic fracture of the sample occurred, and the equilibrium section was practically absent in the loading diagram. Thus, according to the data obtained, it was found that the longer the length of the working area of the sample, the earlier the failure occurs.

In the course of the study, deformation diagrams were obtained on the surface of perforated plates, with the help of which it becomes possible to estimate the maximum longitudinal deformations occurring near the hole.

Taking into account the geometry of the staggered arrangement of the holes presented in the study, we can conclude that the two additional holes did not affect the destruction process. This is confirmed by the similar nature of fracture for all perforated plates and the inhomogeneous fields of transverse, shear and longitudinal deformations on the surface presented in the work. Based on the obtained non-uniform fields, it is possible to establish the presence and increase in the concentration of the level of deformations around the hole, to track the growth of a crack and the subsequent destruction of the sample.

To confirm the experimental results of research and to obtain additional information, a numerical simulation of the process of deformation and destruction of perforated plates was carried out. The calculated data obtained confirm the experimental results of the research.

Acknowledgements
The Experimental studies in the framework of the description of the mechanical behavior of structurally inhomogeneous materials were conducted within the State Assignment of the Ministry of Science and Higher Education of the Russian Federation (FSNM-2020-0027).

Results of the experimental study, video system was carried out in Perm National Research Polytechnic University with financial support of grant of President of Russian Federation for government support of young Russian scientists (Grant MK-1222.2020.8).

Results of numerical calculations was carried out in the Perm National Research Polytechnic University with support of the Russian Science Foundation (Project 16-19-00069).

References
[1] Cunningham D, Harries K A and Bell A J 2015 *Engineering Structures* **95** 8-15
[2] Strungar E M and Wildemann V E 2020 *Frattura ed Integrità Strutturale* **53** 406-416
[3] Strungar E M and Lobanov DS 2020 *Frattura ed Integrità Strutturale* **54** 56-65
[4] Strungar E M, Feklistova EV, Babushkin AV and Lobanov D S 2019 *Procedia Structural Integrity* **17** 965–970
[5] Dattoma V and Giancane S 2013 *Composites Part B: Engineering* **47** 283–289
[6] Sypt P Van Der, Chérif M andBois C 2017 *Int J Fatigue* **103** 86-98
[7] Gyu-Hyeong Lim, Bodjona K, Prasanna Raju K, Fielding S, Romanov V and Lessard L 2018 *Composite Structures* **189** 54-60
[8] Vallons K., Adolphs G, Lucas P, Lomov S V and Verpoest I 2014 *Composites Part A: Applied Science and Manufacturing* **56** 272- 279
[9] Bazant Z P and Di Luizo G 2004 *Intern. J. of Solids and Struct* **41** 7209–7240
[10] Wildeman V, Sokolkin Yu and Tashkinov A 1998 The mechanics of inelastic deformation and fracture of composite materials (Moscow: Nauka) 288
[11] Bian T, Guan Z and Liu F 2018 *Archive of applied mechanics* 1-20
[12] Lobanov D S, Staroverov O A and Tretyakov M P 2020 *IOP Conf Series: Materials Science and Engineering* **747** 012120
[13] Tretyakov M P, Lobanov D S and Wildemann V E 2019 *Procedia Structural Integrity* **17** 865–871
[14] Kozlov M V and Sheshenin S V 2016 *Mechanics of Composite Materials* **51 (6)** 695–706
[15] Kumar D, Roy R, Kweon J-H and Choi J-h 2016 *Compos Mater* **23** 397–419
[16] Lee C-S, Kim J-H, Kim S-K and Ryu D-M 2015 *Compos Struct* **121** 406–419
[17] Tsepennikov M V, Strom A A, Povedev I A and Smetannikov O Y 2016 *PNRPU Mechanics Bulletin* **2** 143–158
[18] Tsepennikov M V, Smetannikov O Yu and Pozhevev I A 2015 *Fundamental and applied problems of engineering and technology* **2** 46-53
[19] Aidi B, Case SW 2015 *Compos Mater* **22** 837–855
[20] Guo Z, Zhu H, Li Y, Han X and Wang Z 2016 *Compos Mater* **23** 1209-1218
[21] Zhikharev MV 2017 *Bulletin of SUSU Series "Mechanical Engineering"* **17** 82-90

Modeling the structure of the TiO_2(rutile)/$SrTiO_3$ heterointerface

S A Tikhonov[1, 2], A P Zavjalov[1,3] and D Yu Kosyanov[1]

[1]Far Eastern Federal University, 10 Ajax Bay, Russky Island, Vladivostok 690922, Russian Federation
[2]Kamchatka Branch of the Geophysical Survey of the Russian Academy of Sciences, 9 Piip Boulevard, Petropavlovsk-Kamchatsky, 683023, Russian Federation
[3]Laboratory of Methods of Synchrotron Radiation, Institute of Solid State Chemistry and Mechanochemistry, Siberian Branch of the Russian Academy of Sciences, 18 Kutateladze Street, Novosibirsk 630128, Russian Federation

E-mail: allser@bk.ru

Abstract. The structure of the TiO_2(rutile)/$SrTiO_3$ interface is modeled using the density functional theory method. Based on calculations, the most stable structure of the heterointerface is determined, for which there are noticeable differences in the structure of the bordering TiO_2 and $SrTiO_3$ layers. This determines mainly the ionic nature of interfacial bonding and the metallic nature of the interface, which can contribute to an increase in the number of charge carriers and the formation of a 2D electron gas. In addition, oxygen vacancies and under-coordinated O and Ti atoms are observed on the bordering TiO_2 layer, indicating excess energy of the heterointerface. The results obtained serve as a basis for using theoretical methods to evaluate the effect of dopants on the thermoelectric properties of TiO_2–$SrTiO_3$ biphasic ceramics.

1. Introduction

Currently, most of the energy is produced by heat engines, which leads to a low efficiency of electricity generation due to the dissipation of most of the heat. Therefore, secondary exhaust heat processing is highly promising for energy efficiency improvement. This gives rise to interest in high-temperature thermoelectric materials.

Earlier [1], the authors of this study presented the concept of creating a new thermoelectric material based on a two-phase ceramic TiO_2-$SrTiO_3$ system. Its peculiarity is that a 2D electron gas does not occur as usual in a specially formed thin layer of some substance, but spontaneously on the TiO_2/$SrTiO_3$ heterointerface (modulation doping).

In [2–4], the conditions for the formation of a 2D electron gas at the interface between thin TiO_2(anatase) and $SrTiO_3$ films were studied in detail, but the formation of biphasic high-density TiO_2–$SrTiO_3$ ceramics requires sintering temperatures above 1 000 °C. Under these conditions, a phase transition of titanium dioxide (anatase→rutile) is possible. It is known [1] that the lower boundary of the rutile conduction band is lower than that of strontium titanate. This implies that the formation of the TiO_2 rutile phase does not significantly affect the thermoelectric properties of the TiO_2–$SrTiO_3$ biphasic ceramics, in which a connected network of 2D electron gas must be formed at the crystal boundary. According to the results of calculations [5] by the density functional theory/ projector augmented wave (DFT/PAW) method using three exchange-correlation generalized gradient approximation (GGA) services out of three possible options of TiO_2 and $SrTiO_3$ layer arrangement, only for the rutile interface TiO_2 (100) on $SrTiO_3$ (111), a connection is observed, which correlates with the experimental results of thin films studies [6]. Therefore, in this paper, the structure parameters of four possible options of the interface between the rutile form TiO_2 (100) and $SrTiO_3$ (111) with different layers on the heterointerface are determined (figure 1).

2. Simulation Details

Calculations of energy and electronic structure were performed in the framework of DFT using the Quantum-ESPRESSO package [7]. The PAW approach [8] and the exchange-correlation functional of GGA were used for modeling, Perdew *et al.* [9] (PW91). DFT/PAW method combines simplicity of first-principles pseudopotential and accuracy of linearized augmented planewave method. PAW is an all-electron technique, which enables one to determine potential using all-electron density. That is why PAW proved to be as one of the most robust approximations for electronic structure modelling of crystals. DFT/PAW using the GGA PW91 functional allows to obtain reliable data on the structure and properties of TiO_2/$SrTiO_3$ interface [4].

The single-particle Kohn-Sham wave function was expanded using plane waves with a cutoff energy of 450 eV. When calculating the interface structure, a 10 Å vacuum region was embedded in supercells along the normal to the boundary, which eliminated undesirable interactions between the plate and its periodic images. The total energy was reduced to an accuracy of 2 meV per atom using the above parameters. When optimizing the structure of individual plates and the heterointerface, the convergence criterion was 0.05 eV/Å.

3. Results and discussion

The accuracy of the chosen method was evaluated by performing a series of calculations of the $SrTiO_3$ and TiO_2 bulk energies and electronic structures. The following optimal $SrTiO_3$ and TiO_2 bulk lattice constants were obtained: a = 3.983 Å, b= 4.660 Å and c = 3.003 Å. The calculated lattice constants are 102% of the experimental value for $SrTiO_3$ [10] and 101.5 % for TiO_2 [11]. The band gap of bulk $SrTiO_3$ is calculated to be 1.69 eV, which is close to the theoretical values of 1.78 eV according to Zhang *et al.* [12] and 1.84 eV according to Fabricius *et al.* [13], but significantly less than the experimental value of 3.25 eV [14]. The estimated band gap of the rutile form of titanium oxide is 1.72 eV, which is also less than the experimental value of 3.00 eV [15]. Such differences between the experimental and calculated energies can be attributed to the well-known disadvantage of the DFT method, which consists in underestimating the band gap of semiconductors.

On the surfaces of $SrTiO_3$(111) and the rutile form of TiO_2 (100), two types of endings were taken into account (figure 1). In total, 4 possible interfaces were built within the framework of the above-mentioned orientation relations. The W_{ad} (Table 1) adhesion energy was estimated by analogy with [4, 16] to predict an energetically stable interface. The unrelaxed W_{ad} value is calculated by subtracting the energy of the boundary layers in the equilibrium state from the energy value of the unrelaxed isolated plates. Then, for each isolated plate and heterointerface, the structure was optimized, which allowed obtaining a relaxed W_{ad} value.

Table 1. Unrelaxed and relaxed values of the adhesion energy W_{ad} (in J/m^2) for all the 4 candidate interfaces.

Structure	Unrelaxed	Relaxed
a	0.29	0.33
b	0.07	0.01
c	0.09	0.15
d	1.56	1.44

IOP Conf. Series: Materials Science and Engineering 1093 (2021) 012031 doi:10.1088/1757-899X/1093/1/012031

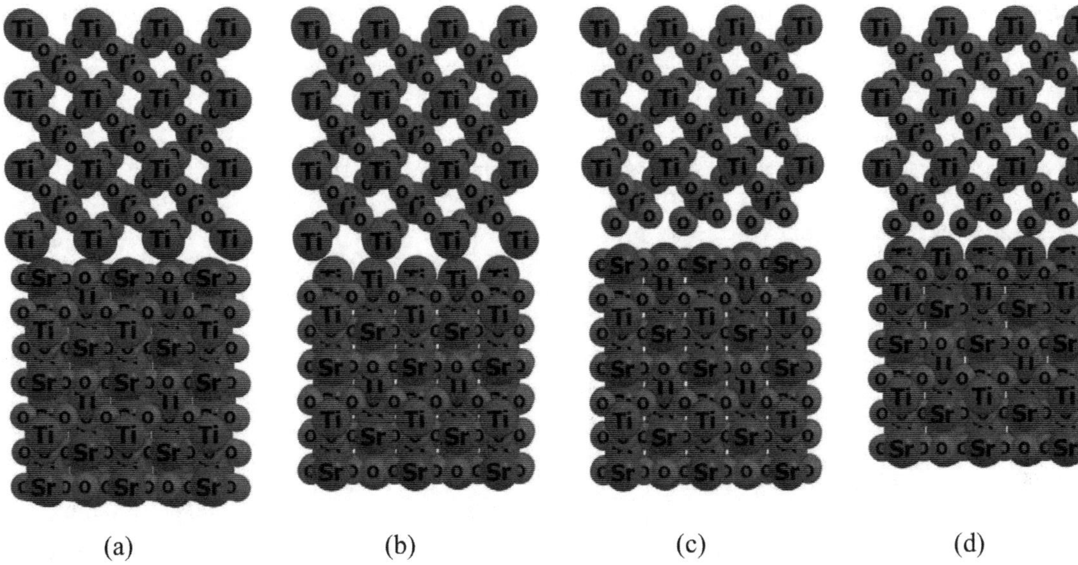

Figure 1. Unrelaxed structures of the heterointerfaces between the rutile form of TiO_2 and $SrTiO_3$.

The highest value of W_{ad} is observed for structure *d* (figures 1 and 2), which can be explained by differences in the structure of the bordering TiO_2 and $SrTiO_3$ layers. This determines the predominantly ionic (with a small amount of covalency) nature of interfacial bonding and the metallic nature of the TiO_2(rutile)/ $SrTiO_3$ interface, which can contribute to the formation of a 2D electron gas on the heterointerface. However, confirmation of this conclusion requires additional experimental studies. In addition, the possibility of forming a 2D electron gas can be influenced by the introduction of various alloying additives in TiO_2 (rutile) or $SrTiO_3$.

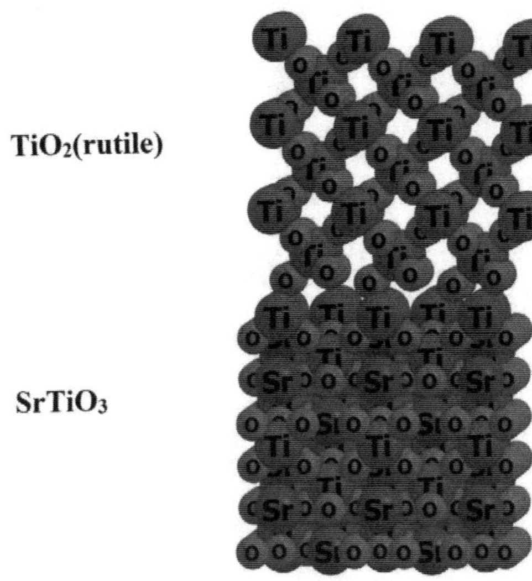

Figure 2. The relaxed structure of the heterointerface (structure *d*) between the rutile form of TiO_2 and $SrTiO_3$.

In the relaxed structure of the TiO_2(rutile)/$SrTiO_3$ heterointerface on the bordering TiO_2 layer, two under-coordinated titanium and oxygen atoms are present in each unit cell, and oxygen vacancies are also observed (figure 2). In addition, in the second TiO_2 layer, some of the Ti atoms are under-coordinated. Additional calculations are required to determine the effect of the number or type of broken bonds on the heterointerface on the W_{ad} value, but it can be assumed that the presence of under-coordinated atoms indicates the excess energy of the heterointerface.

4. Conclusions

Based on DFT calculations, the epitaxial stability of the TiO_2(rutile)/$SrTiO_3$ heterogeneous boundary has been estimated. Modeling has shown the stability of the heterointerface structure, for which the greatest differences are observed in the structure of the bordering TiO_2 and $SrTiO_3$ layers, which determines mainly the ionic nature of interfacial bonding and the metallic nature of the TiO_2(rutile)/ $SrTiO_3$ interface and can contribute to the formation of a 2D electron gas. However, confirmation of this effect requires a series of additional experimental measurements. In any case, it is possible to change the thermoelectric properties of biphasic ceramics TiO_2-$SrTiO_3$ by introducing modifying additives.

After optimizing the structure of the TiO_2(rutile)/$SrTiO_3$ heterointerface, the TiO_2 boundary layer contains under-coordinated Ti and O atoms, as well as oxygen vacancies. This can be attributed to the excess binding energy of the rutile TiO_2 (100) and $SrTiO_3$ (111) layers. The results obtained can be used to calculate the structure of the TiO_2/$SrTiO_3$ interface with various alloying additives in TiO_2 and/or $SrTiO_3$, which helps to predict the optimal amount of the additive and its elemental composition.

Acknowledgements
This work was supported by the Russian Foundation for Basic Research (Project No.18-29-11044).

References

[1] Zavjalov A, Tikhonov S, Kosyanov D 2019 *Materials* **12** 2895
[2] D'Amico N R, Cantele G, Ninno D 2012 *Appl. Phys. Lett.* **101** 141606
[3] Chambers S A, Ohsawa T, Wang C M et al. 2009 *Surf. Sci.* **603** 771–780
[4] Wang Z, Sun R, Chen C et al. 2012 *J. Mater. Sci.* **47** 5148–5157
[5] Xu Z, Salvador P, Kitchin J R 2017 *ACS Appl. Mater. Interfaces* **9** 4106–4118
[6] Burbure N V, Salvador P A, Rohrer G S 2010 *J. Am. Ceram. Soc.* **93** 2530–2533
[7] Giannozzi P, Baroni S, Bonini N et al. 2009 *J. Condens. Matter Phys.* **2139** 395502
[8] Blo¨chl P E 1994 *Phys. Rev. B* **50** 17953
[9] Perdew J P, Chevary J A, Vosko S H, Jackson K A, Pederson M R, Singh D J, Fiolhais C 1992 *Phys. Rev. B* **46** 6671
[10] Howard S A, Yau J K, Anderson H U 1989 *J. Appl. Phys.* **65** 1492
[11] Dewhurst J K, Lowther J E 1996 *Phys. Rev. B* **54** R3673
[12] Zhang C, Wang C L, Li J C, Yang K, Zhang Y F, Wu Q Z 2008 *Mater. Chem. Phys.* **107** 21537
[13] Fabricius G, Peltzer y Blanca E L, Rodriguez C O, Ayala A P, de laPresa P, Lopez Carcı´a A 1997 *Phys. Rev. B* **55** 164
[14] Fujisawa J, Eda T, Hanaya M *Chem. Phys. Lett.* 2017 **685** 23–26
[15] Maheu C, Cardenas L, Puzenat E, Afanasiev P, Geantet C *Phys. Chem. Chem. Phys.* 2018 **20** 25629–25637
[16] Wang Z C, Tsukimoto S, Saito M, Ikuhara Y 2009 *Phys. Rev.* B **79** 045318

The Investigation of the Structure and Properties of Nonwoven Fibrous Materials Based on Polyhydroxybutyrate Biopolymer

P M Tyubaeva[1,2], A K Zykova[1,2], E S Morokov[1], A A Olkhov[1,2,3], V V Podmasteriev[1], A APopov[1,2] and A L Iordanskii[3]

[1]N.M. Emanuel Institute of Biochemical Physics, Russian Academy of Sciences, 4 Kosygin st., Moscow, 119334, Russian Federation
[2]Plekhanov Russian University of Economics, Stremyanny per. 36, Moscow, 117997, Russian Federation
[3]N.N. Semenov Federal Research Center for Chemical Physics, Russian Academy of Sciences, 4 Kosygin st., Moscow, 119334, Russian Federation

E-mail: polina-tyubaeva@yandex.ru

Abstract. Currently, the development and research of nonwoven medical fibrous materials based on biopolymers is an area of a great practical interest. One of the most promising methods for producing nonwoven materials with a highly developed surface is electrospinning (ES). In the article, the possibility of effective sterilization of ultrathin fibers based on polyhydroxybutyrate (PHB) by the ozone treatment was considered. The purpose of this work was to establish the interrelation between changes in the ES process and accompanying changes in the supramolecular structure, morphology, physical and mechanical properties of the fibrous materials based on PHB.

1. Introduction
Highly porous polymer carriers of biologically active substances are widely used in biology and medicine as long-acting matrices, matrices for cell engineering, antibacterial therapeutic systems, controlled release matrices of medical substances, etc. [1]. Fibrous structures and fibrillous materials based on such structures can be obtained in various ways, but the greatest attention should be paid to the method of electrospinning (ES) from a solution of biopolymers. ES has already proven itself as an effective way to produce continuous fibers with a large variation in diameters [2]. Different technical solution of the ES process allow modifying of the fiber surface and controlling of a wide range of properties of the resulting materials [3]. In this work, a number of regularities were established that allow us to vary the properties of fibrous materials based on the biopolymer polyhydroxybutyrate (PHB).

2. Experimental
In the work, nonwoven fibrous materials obtained by the ES method on a single-capillary laboratory unit with the capillary diameter 0.1 mm were used. Molding of the materials was carried out from the forming solutions of the polymer PHB. PHB was produced by the bacteriological synthesis by BIOMER (Germany), grade 16F, viscosity-average molecular weight was 2.6×10^5 Da. In the work, four methods for obtaining the material were used. Methods differed in the formulations of molding solutions with a

content of 7-15% PHB and in the conditions of the ES process. The choice of the conditions varied depending on the properties of the molding solutions: conductivity, viscosity, homogeneity. The voltage was 17-22 kV, the distance between the electrodes was 150-200 mm, the gas pressure on the solution was 10-30 kg(f)/cm^2. Samples of the nonwoven materials with different densities were studied by the electron paramagnetic resonance (EPR) method on the EPR spectrometer EPR-B (Russia). The crystallinity of the materials was studied by differential scanning calorimetry (DSC) on the DSC 214 Polyma of Netzsch (Germany) with a heating rate of 10° K/min. The morphology and the mutual arrangement of fibers in the structure of material was investigated using the optical polarizing microscope Micromed polar-3 (Russia). Microphotographs were prepared in the reflected light at a magnification of 200 times. Mechanical tests were performed on a machine for mechanical analysis DEVOTRANS (Turkey) in compliance with the ASTM D5035-11.

3. Results and discussion

The supramolecular structure of biopolymer fibers obtained by the ES method is highly influenced not only by the molecular characteristics of the polymer, but also by the conditions of its production into the fiber, the curing process, and the stages of subsequent processing [4]. The porosity, geometry, surface topology of each individual fiber, as well as their mutual orientation, make a significant contribution to the formation of physical, mechanical, diffusion and thermal properties of the whole material. It was established, that the supramolecular structure of PHB largely determines the biodegradation rate of the material, its functional quality, which can be improved and controlled in several effective ways: the introduction of modifying additives, chemical processing of fibrous materials, variation of technological parameters of the electroforming process at various stages of fiber production [5, 6].

In the article, two technological solutions for several formulations of forming solutions of the PHB are presented. The considered technological solutions allow obtaining of the fibrillary composites with different fiber morphology, which cause differences in the physical and mechanical characteristics of non-woven materials. This study makes it possible to vary the properties of biopolymer fibrous material, which can be used in products for medicine, veterinary, hygiene, biology, capable of biodegradation after use.

Formulations 1 and 2 allowed obtaining fragile, voluminous, highly porous materials (Figure 1, 1 and 2). It should be noted a large number of defects, thickening, irregularity and heterogeneous of the fibers. Microphotography clearly shows ordered formations in the structure of fibers, which can be obtained because of orientation extraction of the polymer, which occurs during ES. Formulations 3 and 4 allowed getting a fundamentally different structure of the material. The fibrous material could be characterized as a thin and flexible polymer film. At the same time, microphotographs (Figure 1, 3 and 4) show highly developed fibrous structure. It was mentioned, that the average diameter of the fibers decreases and the inter-fiber space increases. Formula 4 was characterized by the highest straightening of fibers and the smallest number of defects.

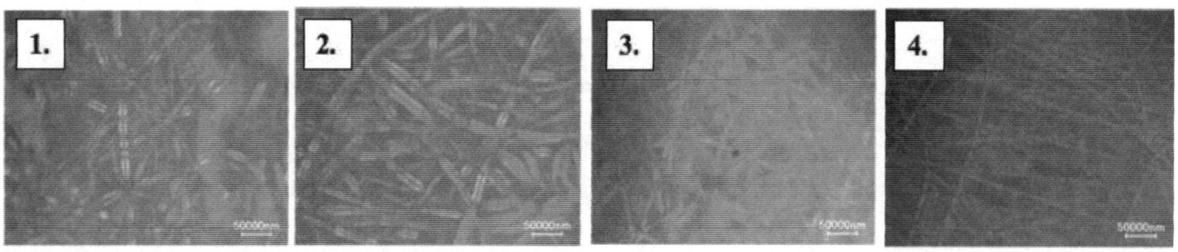

Figure 1. Microphotographs of the nonwoven materials obtained by the ES from the forming solutions based on PHB: 1, 3 – 7% of PHB, 2, 4 – 15% of PHB.

Thus, from Figure 1 it can be seen that the content of PHB in the molding solution did not principally affect on the structure of the whole material. The main factor, which specified the structure type, was based on the conditions of the ES process. In any case, the nature of the dispersion of the diameters of the formed fibers, the number of defects, which include thickening, drops, gluing, breaks, and other current structural features were depending on the formulations. Moreover, these characteristics significantly affected the physical and mechanical properties of the material, which are shown in the Table 1.

The influence of process conditions on the type of structure was established: texturized highly porous or film-like. It was found that voltage, pressure, distance between the electrodes, type of coating of the lower electrode were the key factors. Identical conditions for formulations 1 and 2 contributed to an intensive molding, low jet injection speed, and a smaller diameter of the fibrous trajectory. As a result, the jet was already cured in the material layer. The conditions for formulations 3 and 4 contributed to a less intensive molding. The trajectory of the fiber had a larger diameter and was more stable. The fibers within the layer were already well cured, the surface of the electrode contributed to the denser packing of fibers. All this properties of material samples gives the possibility to vary the average diameters, their physical and mechanical properties and etc. These conditions allowed effective controlling of the quality of the material. Depending on the purpose and field of application, it is possible to obtain a material with different values of porosity, density, surface area and therefore permeability.

Table 1. Results of the physical and mechanical tests of the samples with different structure parameters.

Number of the compound	Physical and mechanical characteristics		Structural characteristics	
	Maximum Strength [N] (Δ±0.5)	Relative Deformation [%] (Δ ±0.2)	Average fiber diameter [μm]	Average thickness of the sample [mm]
1	2.4	4.4	13	0.3
2	0.9	3.3	13	0.2
3	7.8	1.3	8.2	0.16
4	2.3	5.0	9.5	less than 0.1

In view of the fact that the issue of sterilization preparation of these materials before use in medical devices is acute. The highly developed surface makes it difficult to sterilize nonwoven materials by UV treatment. In addition, various studies have shown an uncontrolled decrease in the physical and mechanical properties of the materials based on the PHB after irradiation at a wavelength of 254 nm [7]. PHB is characterized by the temperature of the beginning of destruction of 140-150 °C, which is 40-30 °C below the recommended temperature required for the death of pathogenic microorganisms. This fact makes high-temperature annealing inapplicable for nonwoven materials. The problem of sterilization can be solved by an effective method of disinfection of medical devices, which is the ozone treatment [8]. The effect of the ozone on the supramolecular structure of PHB fibers obtained by different formulations was investigated in this work.

Table 2. Physical-mechanical properties of nonwoven materials with high fibre dispersity before and after ozone treatment (duration of the ozone treatment was 5 minutes long).

Material	Maximum Strength [N]	Relative Deformation [%] (Δ ±0.2)
PHB Initial	1,7	3,4
PHB Ozonized	3,5	7,6

It was found that ozone could not only provide effective sterilization of the material, but also could significantly improve the physical and mechanical parameters. Moreover, the increase in breaking stress and elongation, compared to the initial one, is fixed at the first minutes of ozonation and reaches a maximum in the range of 3-5 minutes of material processing. Table 2 shows the results of the study for one of the formulations. It is important to note that this effect applies to all formulations of PHB-based fibrous materials.

In the work, the reasons for the observed effects are suggested and they consist in the fracture of macromolecules in the amorphous phase followed by more regular laying of macrocycles of PHB. It is important to stress, that at the moment of fiber formation, polymer macromolecules are subjected to orientation and extraction under the action of a complex of physical forces. This leads to structural stresses, as well as disequilibrium in the crystalline and amorphous phases of the PHB. The EPR method allowed seeing the effect of ozone on the amorphous phase of the polymer, where the reaction is quite intense. In the first minutes of ozonation, the correlation time of the probe increased from 30 x 10^{-10} to 60 x 10^{-10} s, which indicated changes in the amorphous phase of the PHB and its densification. The DSC method showed accompanying increases in the degree of crystallinity, as well as an increase in the melting temperature in the samples as they were ozonized during 3-5 minutes. This may indicate that the most stressed sections of the polymer chain are broken, as well as the reorganization of the amorphous phase sections, which contribute to a more orderly arrangement of macromolecules in the crystal regions. These phenomena in the macromolecular structure, most likely, lead to an increase in physical and mechanical properties in the first minutes of ozonation. Moreover, if the ozonation time exceeds 7-8 minutes, then the destruction of the material is visible, observed in a uniform drop in the correlation time below the initial one.

4. Conclusions

Different types of materials could be obtanied within different conditions of the electrospinning process. The study of the physical and chemical properties of such nonwoven materials obtained from polymer solutions after their effective sterilization is an area of a great interest. The article established and showed an example of an effective method of sterilization that can improve the physical and mechanical characteristics of the material, and also considered the reasons for this process, due to the influence of ozone on the supramolecular structure of the material.

Acknowledgments

This work was supported by Russian Foundation for Basic Research (project no. 20-33-70095). The study was carried out using scientific equipment of the Center of Shared Usage «New Materials and Technologies» of Emanuel Institute of Biochemical Physics and the Common Use Centre of Plekhanov Russian University of Economics.

References

[1] Xie G, Wang Y, Han X, Gong Y, Wang J *et al* 2016 Ind. Eng. Chem. Res. **55** № 26 pp 7116
[2] Filatov Yu N 1997 Electric Formation of Fibrous Materials pp 10-83
[3] Zhang C X, Yuan X Y, Wu L L *et al* 2005 *Eur. Poly. J. V.* **41** № 3 pp 423-32
[4] Deitzel J M, Beck Tan N C, Kleinmeyer J D *et al* 1999 *Army Research Laboratory Technical Report* pp 2
[5] Olkhov A A, Tyubaeva P M, Staroverova O V *et al* 2016 *AIP Conf. Proc.* **1736** pp 020098-020102
[6] Tyubaeva P M, Olkhov A A and Podmasteriev V V 2018 *Journal of Physics: Conference Series.* pp 012034-012038
[7] Odelius K, Plikk P and Albertsson A-C 2008 *Biomaterials* **29** № 2 pp 129–140
[8] Schiffman J D and Schauer C A 2008 *Polymer Reviews* **48** №2 pp 317-352

Structure, morphology and mechanical properties of arrays of zirconia nanofibers at different heat treatment modes

A I Tyurin, V V Rodaev, V V Korenkov, S S Razlivalova, A O Zhigachev, V M Vasyukov and T S Pirozhkova

Nanotechnology and Nanomaterials Research Institute, G.R. Derzhavin Tambov State University, 33, Internatsionalnaya str., Tambov, 392000, Russia

E-mail: tyurin@tsu.tmb.ru

Abstract. Mats of yttria-stabilized tetragonal zirconia nanofibers were prepared in the present research. The effect of the mats calcination temperatures (600, 900, and 1200 °C) on their structure and morphology was investigated. It was found that phase composition of the fibers did not change in all range of the calcination temperatures, while the average grain size in fiber increased from 8 to 39 nm. Multi-cyclic and single loading-unloading nanoindentation testing of the ceramic mats showed that the hysteresis loop energy in samples decreases with higher calcination temperature. Hardness and the elastic modulus measured by spherical (with 10 and 250 μm radii) and sharp Berkovich indenters were the highest in the mats calcined at 900 °C. This calcination temperature can be considered an optimal one for preparation of nanofibers mats which mechanical properties are of importance for their application.

1. Introduction

It's well known that electrospinning is the most commonly used technique for a large-scale nanofibers production due to its simplicity, versatility, high reproducibility and also the ability to manufacture filaments with a desired diameter, orientation, composition and morphology [1]. Polymer, composite, ceramic, metallic, carbon functional nanofibers can be produced using electrospinning [2]. Among ceramic nanofibers, ZrO_2 filaments occupy a special place due to their excellent thermal and chemical stability, oxygen-ionic conductivity, polymorphism, and amphoteric surface properties. Along with a high aspect ratio and a large specific surface area, this makes ZrO_2 nanofibers promising for bone tissue regeneration [3], solid oxide fuel cells [4], protection from electromagnetic interference [5], etc. While most zirconia nanofiber mats are functional, there is always a requirement for their mechanical properties. A nonwoven material containing spontaneously oriented ceramic nanofibers is a porous non-homogenized medium. The main local mechanical properties of the medium, such as hardness and elasticity, are up to several tens of MPa. Such values of hardness are typical for soft polymers [6]; they are well below ones reported for even highly porous ceramics [7]. The same is true for the elastic modulus [8].

Therefore, overcoming unnecessary fragility and improving the flexibility of ceramic fiber mats comes to the fore. It is well known that the greatest commercial value is 3Y-TZP bulk ceramics, which is obtained by adding 3 mol% Y_2O_3 as a stabilizer agent in ZrO_2 and is a fully tetragonal fine-grained ceramic that is characterized by the best combination of hardness, fracture toughness, bending strength and Young's modulus [9]. But only bulk 3Y-TZP ceramics possess such outstanding characteristics. Typically, the Young's modulus of a 3Y-TZP nanofiber mat does not exceed units of MPa, which was significantly lower than the typical value of 210 GPa for bulk 3Y-ZrO2 ceramics. This indicates 3Y-TZP nanofiber mat remarkable flexibility compared to traditional bulk 3Y-TZP ceramics. But other mechanical properties of the mats, such as hardness and elastic modulus were 4-5 orders of magnitude less than in bulk ceramics. Some autors [3, 10, 11] observed that the increase in annealing temperature led, first, to the partial loss of the ceramic mat flexibility and then to its fragility, but it has not been studied in detail. To study the mechanical properties of 3Y-TZP ceramic mats, we used nanoindentation [12]. We used the methods of single and multi-cycle loading with indenters of different sharpness and shape (spherical and pyramidal) to study the response of 3Y-TZP mats to such a load. The aim of this work was a comprehensive investigation of possible causes for the evolution of the mechanical strength and flexibility of ceramic mats made of nanofibers during nanoindentation with blunt (spherical) and sharp (pyramidal) indenters under the action of the same load and different annealing temperatures.

2. Materials and methods

3Y-TZP nanofiber mats were prepared from a solution of polyacrylonitrile (PAN, molecular weight Mw = 150,000, Sigma-Aldrich, USA) in N, N-dimethylformamide (DMF, Sigma-Aldrich, USA) followed by the addition of zirconium acetylacetonate (ZrAA, Sigma- Aldrich, USA) and yttrium nitrate hexahydrate (Sigma-Aldrich, USA) after stirred at 80 °C until the solution became transparent. Details of the preparation of ceramic nanofiber mats are described in [13]. The prepared solution was electrospun using electrospinning machine NANON-01A (MECC, Japan). The following operative parameters were chosen to fabricate smooth nanofibers: the accelerating voltage was 22 kV, the distance between the needle tip and the flat collector covered by aluminum foil was 20 cm. As-spun mats were calcined at temperatures 600, 900 and 1200 °C in air atmosphere.

The mechanical properties of fabricated zirconia nanofiber mats were measured by the nanoindentation method [14-19] using NanoIndenter® G200 (MTS Nano Instruments, USA) and TI-950 (Hysitron, USA) nanoindentometers. The nanoindenters operated in displacement control mode with a 0.05 s^{-1} strain rate target and a 10 s hold at the maximum depth. Reduced elastic modulus and nanohardness were measured by diamond and zirconia spherical indenters with nominal radii of 10 μm and 250 μm, respectively. A sharper Berkovich indenter, the tip of which has R = 20 nm, was also used in this work.

The studies were carried out in a multi-cycle test mode [20-22], which makes it possible to effectively evaluate the properties of a ceramic fiber mat at various contact scales. In the process of multi-cycle testing, 5 repeated cycles loading and unloading are localized in the same indentation. The load on the indenter was proportionally varied from 0 to 5 mN. In this case, the residual load should keep the indenter in contact with the sample without side slip. At the end of unloading, a 20 s hold was provided to correct the thermal drift. To minimize thermal drift prior to testing, all tests were set to an initial drift rate of 0.08 nm/s. All the values were averaged over 10 measurements.

3. Results and discussion

Electrospun composite nanofibers are randomly distributed to form a nonwoven mat. They are initially cylindrical with a smooth surface. The average diameter of the produced filaments is 575 ± 65 nm. After calcinations at 600 °C the average diameter of the fibers reduced to 162 ± 15 nm due to ZrAA decomposition and PAN removal. Their surface remained smooth despite the relatively high temperature. The appearance and growth of ZrO_2 grains begins at temperatures above 650 °C that results in appearance of a rough surface. A further increase in the calcination temperature to 1200 °C is accompanied, on the one hand, by further growth of ZrO_2 grains, and on the other hand, leads to a slight shrinkage of the fibers due to sintering. As a result of these processes, the average fiber diameter increases to 145 ± 16 nm, and their surface becomes even more rough.

The rise in fibers calcination temperature affects their specific surface area and porosity too. From nitrogen adsorption-desorption isotherms at 77 K of zirconia fibers prepared at different temperatures (figure 1) it can be seen that increase in calcination temperature results in hysteresis loop narrowing and the quantity of adsorbed nitrogen decrease. It indicates zirconia fibers porosity reduction with increase in calcination temperature due to ZrO_2 grain growth and sintering. In our case, pores are formed by the boundaries of adjacent ZrO_2 grains. These grain growths are also causes zirconia fibers specific surface area decrease with rise in calcination temperature.

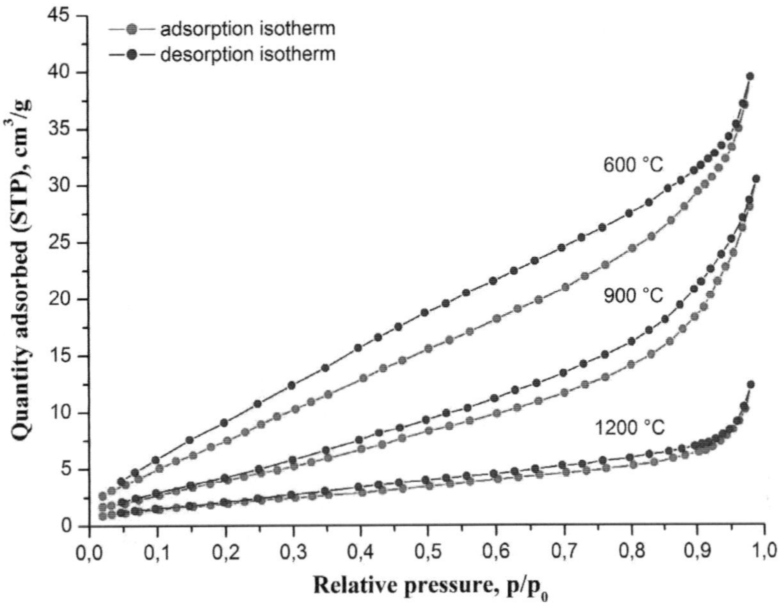

Figure 1. Nitrogen adsorption-desorption isotherms at 77 K of electrospun ZrAA/$Y(NO_3)_3$/PAN fibers calcined at 600 °C, 900 °C and 1200 °C.

The evolution of the grain structure of nanofibers with increasing calcination temperature is confirmed by XRD data. We saw that the fibers calcined at 600 ° C have a crystalline structure and all the observed reflections correspond to t-ZrO_2. This indicates that these fibers are composed only of t-ZrO_2 grains. At a calcination temperature of 900 °C, the average grain size of t-ZrO_2 was 18 nm, and at a temperature of 1200 °C it reached 39 nm. No additional reflections from other 3Y-ZrO_2 phases are observed in the X-ray diffraction patterns with an increase in the calcination temperature. This is evidence that heat treatment of nanofibers in the range 600–1200 °C does not affect the phase composition of the fibers, although the average grain size of t-ZrO_2 increases by almost 5 times.

Two parameters can be relatively easily extracted from instrumental indentation tests: hardness *H* and Young's modulus *E*, representing the plastic and elastic properties of the material, respectively. Both of these quantities depend in a rather complicated way on the testing conditions, in particular on the geometry of the indenter [23]. Typically, nanoindentation provides results from a locally homogenized volume at a micro- or nanoscale. It should be noted that the use of nanoindentation in the case of heterogeneous and porous materials, such as ceramic fiber mats, is very limited. This limitation is due to the fact that the theory is based on the assumption of a homogeneous and isotropic half-space without scale restrictions, and the data obtained are average values depending on the type of indenter and the depth of indentation. The general behavior of 3Y-TZP mats during nanoindentation directly depends on the formation of individual phases and their microstructures and is associated with the possible mixing of unreacted components or the occurrence of chemical reactions that develop after such mixing [10].

Figure 2 illustrates depth-dependencies of the elastic modulus and hardness of the 3Y-TZP nanofibers mats tested by indenters with different sharpness and shape. As we seen on figure 2(*b*, *c*), the effect of the calcination temperature on the elastic modulus is non-monotonous, with the highest values observed for the mats calcined at 900 °C. At this calcination temperature, the average size of tetragonal grains usually does not exceed 18 nm, and with a further increase in temperature to 1200 °C they reach 39 nm in diameter. Such a rearrangement of the grain structure is accompanied by a simultaneous decrease in the porosity of nanofibers. It should be noted also that the elastic modulus obtained at calcined 1200 °C is equal $E = 0.55 \pm 0.06$ MPa is below the typically reported values [3, 10, 11]. In this case, E is lower, and the elasticity of the mats at this annealing temperature is even higher than at $T = 600$ C, and this is observed for both spherical indenters, the radii of which differ by a factor of 25. We also see that the absolute values of hardness for a spherical tip with $R = 250$ μm differ by 2 times from the higher values obtained with a 10 μm indenter tip. This is because a blunt spherical indenter has a higher contact surface, allowing more material to be integrated under the same load.

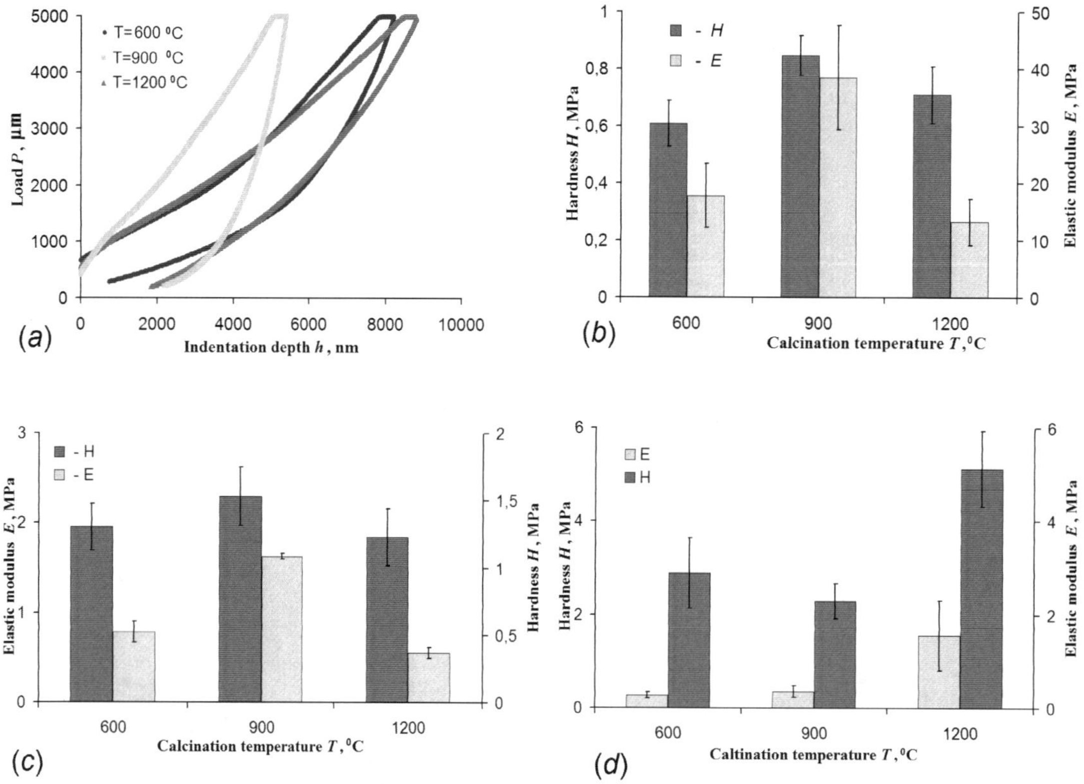

Figure 2. *P-h* diagram (*a*), elastic modulus and hardness of a mat of ceramic 3Y-TZP nanofibers under indentation for different annealing temperatures by: (b) spherical indenter $R = 250$ μm, (c) spherical indenter $R = 10$ μm, (d) Berkovich indenter $R = 20$ nm.

Berkovich pyramidal indenter is much sharper than spherical indenters. The spherical blunting tip of such an indenter does not exceed 20 nm. Figure 2*d* shows how the elastic modulus and hardness change as a function of the calcination temperature during nanoindentation with a Berkovich tip. All the regularities of the indentation scale effect (ISE) in hardness are also fulfilled here: the sharper the indenter used, the greater the obtained hardness value. The elastic modulus has the highest value equal to 1.56 ± 0.65 MPa for $T = 1200$ °C.

From the energetic point of view, the method of multi-cycling loading is most suitable for indentation of ceramic fiber mats. As seen in figure 3, multi-cycling load-unload-reload-unload tests contain both unloading and reloading curves. The beginning of the reloading curve has a lesser effect on the formation of material pop-in around the indentation, that contributes to a stricter observance of the Oliver-Pharr conditions [14]. We see that the loop energy (the loop area) decreases with the increase in calcination temperature. In the last load-unload cycle on the mats calcined at 1200 °C the hysteresis loop energy was approximately 2.5 times lower than the one registered in the 600 °C calcined mats. It was found earlier that the hysteresis observed under multicycle nanoindentation may appear because of quasi-brittle fracture, phase transformation, viscoelastic or viscoelastic deformations. It was experimentally proved that all these processes increase the area of the hysteresis loop. But as we have seen in figure 4, at the present research the hysteresis loops narrow with the increasing calcination temperature. In our case it can be assumed that increase in crystallinity of the fibers and growth of the average grain size may serve as such mechanism.

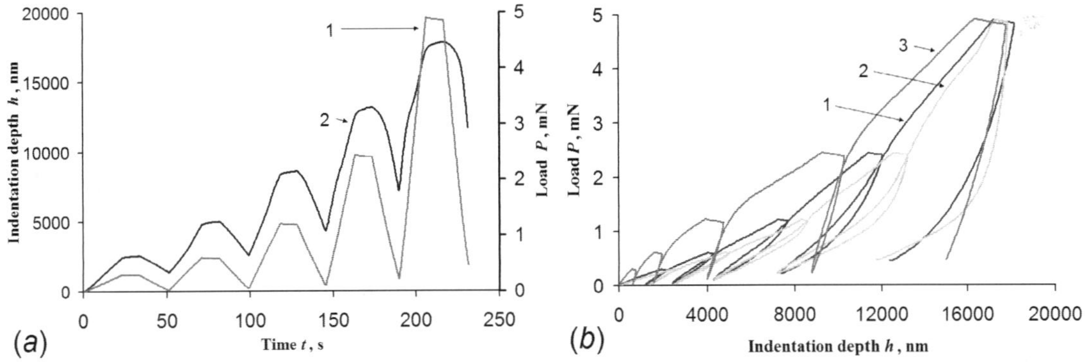

Figure 3. (*a*) - dependence of the load (1) and the indentation depth (2) on the time of the multi-cycle test; (b) - *Ph*-diagrams of 5 loading-unloading cycles at different annealing temperatures: 1 - 600 °C, 2 - 900 °C, 3 - 1200 °C.

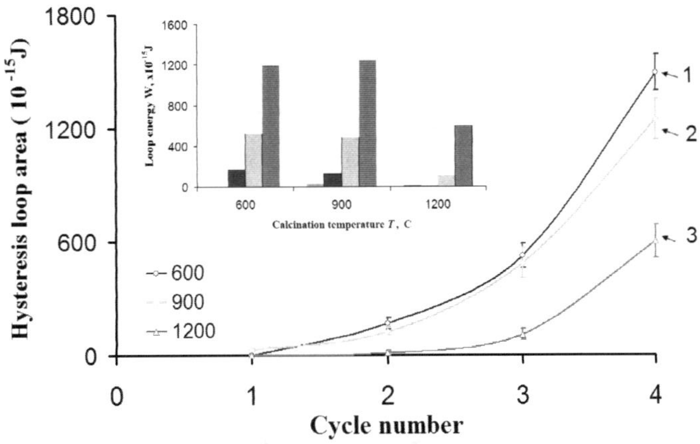

Figure 4. Dependence of the energy of the hysteresis loop on the annealing temperature of ceramic fiber mats: 1 – 600 °C, 2 – 900 °C, 3 – 1200 °C.

The highest mechanical properties are achieved in mats calcined at 900 °C. It was previously shown that at this temperature, micro-roughness begins to appear on the surface of nanofibers due to grain growth, which affects the elasticity and hardness of the mats.

4. Conclusion

The mats of ceramic nanofibers were prepared from solution of PAN in N,N-dimethylformamide with addition of zirconium acetylacetonate and yttrium nitrate hexahydrate by the electrospinning technique. Their were calcined at 600, 900, and 1200 °C; the resulting fibers had a diameter of 145±16 nm. The phase composition of the mats at these temperatures was 100 % tetragonal zirconia phase. It was found that the average grain size increases from 8 to 39 nm with an increase in the calcination temperature to 1200 °C.

Some mechanical properties of the mats were determined by the nanoindentation method. Experiments with indenters of different sharpness have shown that mats calcined at 900 °C have the highest hardness and modulus of elasticity. Optimal mechanical properties are believed to occur at a calcined temperature of 900 °C due to the balance between sufficient grain growth and little or no fiber cross-sintering. This annealing temperature can be considered optimal for the manufacture of nanofiber mats, the mechanical properties of which are important for their application.

Acknowledgments

The research was financially supported by the Russian Fund for Basic Research: projects № 18-29-17047 (synthesis, study of the microstructure and of the mechanical properties) and № 19-03-00634 (investigation of the phase composition), and by TSU named after G.R. Derzhavin, order No. 591-3 of February 25, 2020 (nanoindentation tests). The research was carried out with the use of equipment of G.R. Derzhavin Tambov State University Common Use Center.

References

[1] Xue J, Wu T, Dai Y and Xia Y 2019 *Chem. Rev.* **119** 5298
[2] Thenmozhi S, Dharmaraj N, Kadirvelu K and Kim H Y 2017 *Mat. Sci. Eng. B.* **217** 36
[3] Gazquez G C, Chen H, Veldhuis S A, Solmaz A *et al* 2016 *ACS Nano* **10** 5789
[4] Koo J Y, Lim Y, Kim Y B, Byun D and Lee W 2017 *Int. J. Hydrogen. Energ.* **42** 15903
[5] Im J S, Kim J G, Bae T-S and Lee Y-S 2011 *J. Phys. Chem. Solids* **72** 1175
[6] Deuschle J, Enders S and Arzt E 2007 *J. Mater. Res.* **22** 3107
[7] Chen Z, Wang X, Atkinson A and Brandon N 2016 *J. Eur. Ceram. Soc.* **36** 1435
[8] Wei J, McFarlin B L and Vagoner Johnson A J 2016 *J. Mater. Res.* **31** 2672
[9] Gautam C, Joyner J, Gautam A, Raoc J and Vajtai R 2016 *Dalton Trans.* **45** 19194
[10] Castkova K, Maca K, Sekaninova J, Nemcovsky J and Cihlar J 2017 *Ceram. Int.* **43** 7581
[11] Stanishevsky A, Yager R, Tomaszewska *et al* 2019 *Ceram. Inter.* **45** 18672
[12] Lucca D A, Herrmann K and Klopfstein M J 2010 *CIRP Annals - Manufacturing Technology* **59** 803
[13] Rodaev V V, Razlivalova S S, Zhigachev A O *et al* 2019 *Polymers* **11** 1067
[14] Oliver W C and Pharr G M 1992 *J. Mater. Res.* **7** 1564
[15] Golovin Yu I, Tyurin A I, Aslanyan E G, Pirozhkova T S, Vorobev M O 2016 *Measurement Techniques* **59**(9) p. 911-915
[16] Golovin Yu I, Tyurin A I, Aslanyan E G, Pirozhkova T S, Vasyukov V M 2017 *Physics of the Solid State* **59**(9) p. 1803-1811
[17] Golovin Yu I, Iunin Yu L, Tyurin A I 2003 *Doklady Physics* **48**(9) p. 505-508
[18] Golovin Yu I, Tyurin A I 2000 *Physics of the Solid State* **42**(10) p. 1865-1867
[19] Golovin Yu I, Tyurin A I, Khlebnikov V V 2005 *Technical Physics* **50**(4) p. 479-483
[20] Chen J and Bull S J 2008 *J. Phys. D: Appl. Phys.* **41** 074009
[21] Richter A, Wolf B, Nowicki M, Smith R, Usov I O, Valdez J A and Sickafus K 2006 *J. Phys. D: Appl. Phys.* **39** 3342
[22] Huang Q Q, Song Y, Liu W W, Chen Y, Qi F, Zhao D and Wang Y G 2015 *Ceramics International* **41** 12349
[23] Králík V and Němeček J 2014 *Mater. Sci. Eng. A* **618** 118

Features of reactive SPS of $SrTiO_3$–TiO_2 biphasic ceramics

A P Zavjalov[1,2], O O Shichalin[1,3], S A Tikhonov[1,4] and D Yu Kosyanov[1]

[1]Far Eastern Federal University, 10 Ajax Bay, Russky Island, Vladivostok 690922, Russian Federation
[2]Laboratory of Methods of Synchrotron Radiation, Institute of Solid State Chemistry and Mechanochemistry, Siberian Branch of the Russian Academy of Sciences, 18 Kutateladze Street, Novosibirsk 630128, Russian Federation
[3]Institute of Chemistry, Far-Eastern Branch of the Russian Academy of Sciences, 159 100-let Vladivostoku Avenue, Vladivostok 690022, Russian Federation
[4]Kamchatka Branch of the Geophysical Survey of the Russian Academy of Sciences, 9 Piip Boulevard, Petropavlovsk-Kamchatsky, 683023, Russian Federation

E-mail: Zav_Alexey@list.ru

Abstract. $SrTiO_3$ and biphasic $SrTiO_3$–TiO_2 ceramics were obtained by reactive spark plasma sintering (SPS) of mixtures of strontium carbonate $SrCO_3$ and titanium dioxide TiO_2. For monophasic $SrTiO_3$, there is a linear seal of the mixture in the heating process. Biphasic ceramics at a temperature of about 1320°C are characterized by eutectic melting of the $SrTiO_3$–TiO_2 mixture, which allowed estimating the temperature gradient in the system ~77°/sm. Microstructural studies have confirmed the formation of biphasic $SrTiO_3$–TiO_2 ceramics with a homogeneous structure and a grain size of 1÷2 μm.

1. Introduction

A significant share of the electricity received by mankind is limited by the performance of heat engines, which leads to energy losses in the form of dissipated heat. The efficiency of power plants can be increased by directly consuming waste heat, for example, for heating. However, it is important to be able to get additional electricity. Direct conversion of heat to electricity is possible using thermoelectric materials, which are used, for example, in systems with radioisotope thermoelectric generators to ensure autonomous operation of satellites, beacons, etc.

The key characteristic of these materials is the thermoelectric q-factor *ZT*. At relatively low temperatures (up to 300°C), the $ZT \sim 1$ value is typical for widely used materials based on Si–Ge and Bi–Sb–Te [1, 2]. At temperatures of 500–1000°C, high *ZT* values (up to ~1.5) are achieved using compounds with heavy metals [2], but at higher temperatures (≥1000°C), these materials may be subject to thermal degradation. Therefore, more stable oxide compounds should be considered for use as high-temperature thermoelectrics.

Due to the features of the state density function, low-dimensional electronic systems have high thermoelectric characteristics [3]. Based on data for thin Sr layers $Sr(Ti_{0.8}Nb_{0.2})O_3$ in the amount of $SrTiO_3$ [4] and other reported parameter values, the paper [5], using numerical modeling, showed that the structure consisting of cubic grains $(Sr_{1-x}La_x)TiO_3$, 16 unit cells in size, separated by $Sr(Ti_{0.8}Nb_{0.2})O_3$ layers in 1 unit cell, could reach *ZT* values ~1.15–1.2 at room temperature. This is achieved due to the properties of 2D electron gas (2DEG) in thin layers, as well as their influence on the transport properties of the entire material. Creating such a structure requires the use of complex nanoconstruction techniques. In [6], cubic particles of this composition were obtained, but they have a much larger size, and it is not clear how to consolidate them without destroying the thin layers.

However, [4] demonstrated the spontaneous occurrence of 2DEG at the interface between the $SrTiO_3$ and TiO_2 layers. Therefore, in the review [7], the authors proposed a biphasic $SrTiO_3$–TiO_2 ceramics with a staggered grain structure of the components. The phase interleaving must form a connected 2DEG network that permeates the entire material. A similar structure of dense ceramic grains was obtained earlier for the Y_2O_3–MgO system using the SPS method [8]. This paper presents the features of obtaining biphasic $SrTiO_3$–TiO_2 ceramics (50:50 for vol.%) by reactive SPS of a powder mixture of $SrCO_3$ and TiO_2, as well as monophasic SrTiO3 ceramics for comparison from a powder mixture of the same components (1:1 by mole).

2. Experimental section

Commercial powders (Sigma-Aldrich) of titanium dioxide TiO_2 (anatase, 99.8%) and strontium carbonate $SrCO_3$ (99.9%) were used to prepare two types of mixtures. The first type is designed to produce $SrTiO_3$ ceramics by the reaction $SrCO_3 + TiO_2 \rightarrow SrTiO_3 + CO_2\uparrow$. The second type has an excess of TiO_2 to produce biphasic $SrTiO_3$–TiO_2 ceramics (50:50 for vol.%) taking into account the phase transformation anatase→rutile during SPS. The mixtures were prepared by grinding in a Pulverisette 6 planetary mill (Fritsch, Germany) in an isopropanol at 280 rpm for 10 hours, followed by drying, granulation, and calcination steps to remove organic components. Reactive SPS experiments were performed on the SPS-515S (Dr. Sinter Lab, Japan) in a graphite crucible Ø15 mm (sample 2 g) in a vacuum (~6 Pa) at a mechanical load of ~21.5 MPa and isothermal exposure at 1000÷1400°C for 5 min.

The microstructure and elemental analysis of the samples were was analyzed by the scanning electron microscope (SEM) Carl Zeiss Ultra 55+ (Carl Zeiss, Germany) with an energy-dispersive X-ray (EDX) detector X-Max (Oxford Instruments, UK). The phase composition was performed using a powder X-ray diffractometer D8 Advance (Bruker, Germany) using $CuK_{\alpha1,2}$ doublets and a range of double reflection angles from 5° to 90° with a step of 0.02° and a speed of 0.5°/min.

3. Results and Discussion

Figure 1 shows the densification curves and the temperature modes of the obtained ceramic samples. Attention should be paid to the stage of linear densification of $SrTiO_3$ ceramics (Figure 1a) (in the first ~10 minutes of the experiment), which is not observed when obtaining biphasic $SrTiO_3$–TiO_2 ceramics (Figure 1b). This difference can be explained by the higher concentration of $SrCO_3$ in the mixture for the preparation of monophasic $SrTiO_3$ ceramics since strontium carbonate is much softer than titanium dioxide. That is, when sintering monophasic ceramics, due to the deformation of the $SrCO_3$ phase, the mixture is compacted as it is heated. For biphasic $SrTiO_3$–TiO_2 ceramics, the TiO_2 phase forms a stronger frame and does not allow such deformation.

In addition, for both mixtures, it can be noted that exposure at 1000°C does not lead to significant densification. At higher temperatures, the nature of the seal changes significantly; however, significant densification occurs at temperatures from 1000°C to 1100–1200°C and changes slightly at higher temperatures. According to XRD data, the $SrCO_3$ phase is not detected for samples sintered at 1000°C, i.e., the reaction between the components occurs at lower temperatures. Therefore, it can be concluded that non-reactive sintering processes are activated in the temperature range from 1000°C to 1100–1200°C.

For biphasic $SrTiO_3$–TiO_2 ceramics, it was not possible to sinter the mixture at 1400°C, because when the temperature reached ~1320°C, a sharp increase in densification occurred, and the experiment was interrupted. This sintering behavior is usually observed when the material is melted. In this case, the authors associate melting with the fact that there is a eutectic for the $SrTiO_3$–TiO_2 system. Sintering at 1300°C was carried out to the end. However, the resulting sample had a hole in the central part. In addition, a small jump in densification in the ~15th minute of the experiment can be observed.

This result is associated with the heterogeneity in the temperature distribution in the sintered sample. There was a slight overheating in the central region, which led to local destruction of the melting region and was reflected in a jump in densification, after which sintering continued in normal mode. Sintering with exposure at 1200°C did not have the characteristic features of melting. Therefore, this mixture was sintered separately at an intermediate holding temperature of 1250°C, which also had no special features. The melting of the sample at a temperature of ~1320°C and the destruction of the central region at ~1260°C in the experiment with exposure at 1300°C makes it possible to obtain unique information about the SPS process – to estimate the temperature gradient $\Delta T/\Delta r \sim 77$°C/sm.

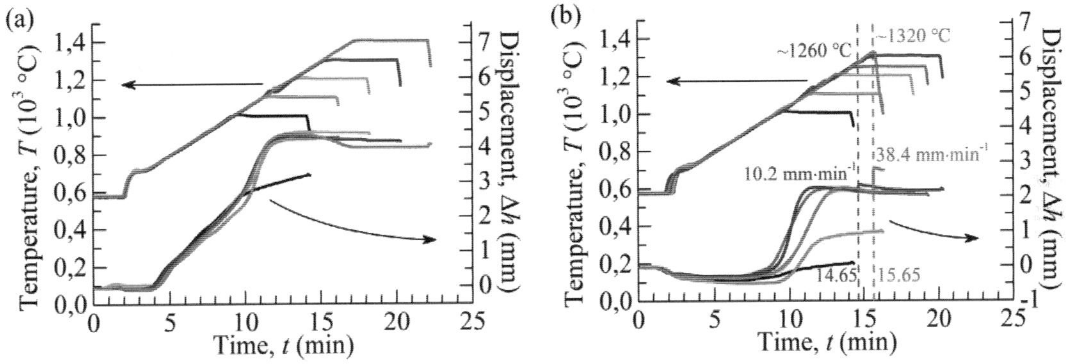

Figure 1. Densification curves and temperature modes for ceramics obtained by reactive SPS of powder mixtures of strontium carbonate $SrCO_3$ and titanium dioxide TiO_2: (a) $SrTiO_3$; (b) $SrTiO_3$–TiO_2 (50:50 for vol.%).

Figure 2 shows X-ray diffractograms of ceramic samples obtained at 1200°C for monophasic and 1250°C for biphasic ceramics. The complete reaction can be seen with the formation of the target phases – tausonite $SrTiO_3$ (lattice constant 3.90481 Å for monophasic and 3.90778 Å for biphasic ceramics) and rutile TiO_2 (lattice constants $a = b = 4.59492$ Å and $c = 2.96147$ Å for biphasic ceramics). No impurity phases were detected. Figure 3 shows the microstructure and elemental analysis data (at.%) of these ceramics. The formation of a homogeneous microstructure with an average grain size of 1÷2 μm is shown. Elemental composition based on the results of EDX studies using the XRD data. Using the EDX method, which provides information averaged over a volume larger than the phase sizes, allowed for local phase identification. In Figure 3b, dark grains correspond to the composition of TiO_2, and light grains correspond to $SrTiO_3$. No impurity elements were detected.

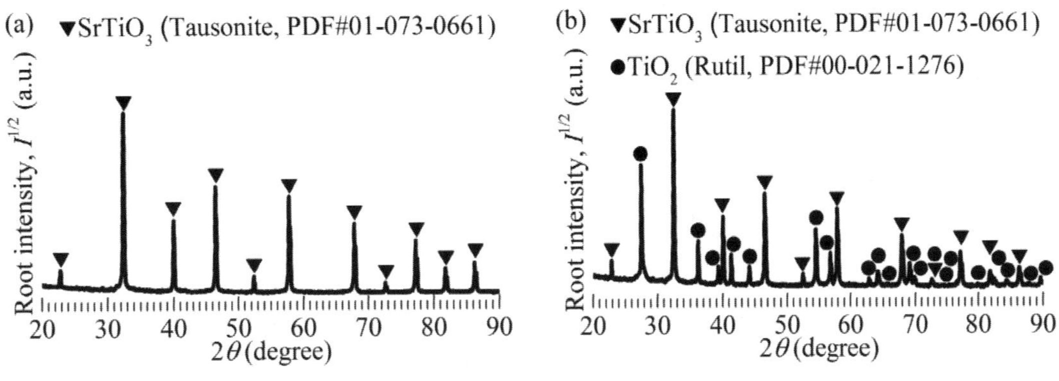

Figure 2. X-ray diffractograms for ceramics obtained by reactive SPS of powder mixtures $SrCO_3$–TiO_2: (a) $SrTiO_3$; (b) $SrTiO_3$–TiO_2 (50:50 for vol.%).

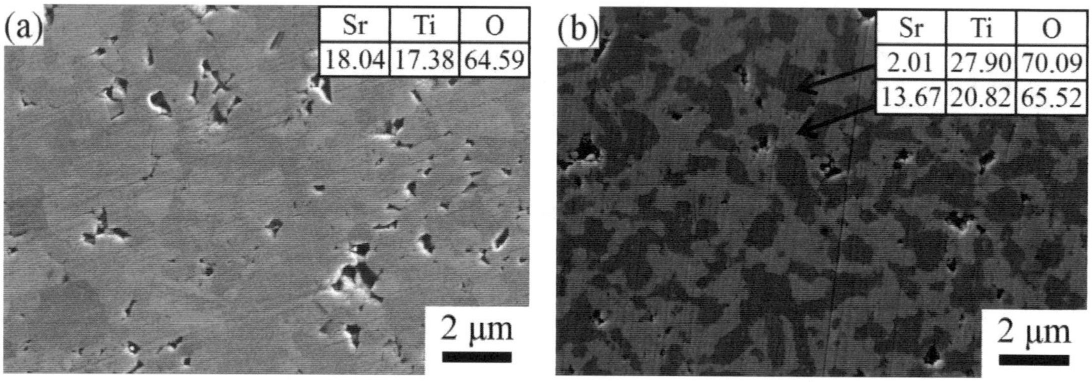

Figure 3. SEM images of the ceramics obtained by reactive SPS of powder mixtures $SrCO_3$–TiO_2: (a) $SrTiO_3$; (b) $SrTiO_3$–TiO_2 (50:50 for vol.%).

4. Conclusions

Two series of experiments were performed to obtain $SrTiO_3$ and biphasic $SrTiO_3$–TiO_2 ceramics from mixtures of $SrCO_3$ and TiO_2 with different component ratios using the reactive SPS method. It has been found that when monophasic ceramics are produced, the mixture is linearly compacted as it is heated, and non-reactive sintering processes are activated from 1000°C to 1100–1200°C. When obtaining biphasic ceramics $SrTiO_3$–TiO_2, eutectic melting is observed when the temperature reaches ~1320°C. Partial melting of $SrTiO_3$–TiO_2 ceramics sintered at a holding temperature of 1300°C allowed estimating the temperature gradient in the system $\Delta T/\Delta r$~77°/sm. The experimental determination of such parameters is unique. The research of the microstructure of obtained samples allowed determining that the reactive SPS results in the formation of monophasic $SrTiO_3$ and biphase $SrTiO_3$–TiO_2 ceramics of a homogeneous structure with an average grain size of 1–2 μm.

Acknowledgments

This work was supported by the Russian Foundation for Basic Research (Project No. 18-29-11044).

References

[1] Nolas G S, Sharp J and Goldsmid J 2001 *Thermoelectrics. Basic Principles and New Materials Developments* (Berlin, Heidelberg: Springer) p 293
[2] Raj B (Eds.), Van de Voorde M (Eds.) and Mahajan Y (Eds.) 2017 *Nanotechnology for Energy Sustainability* (Weinheim, Germany: Wiley-VCH) p 1221
[3] Heremans J P 2005 *Acta Phys. Pol.* A **108** 609–634
[4] Ohta H, Kim S, Mune Y, et al. 2007 *Nature Mater.* **6** 129–134
[5] Zhang R, Wang C and Li J 2010 *J. Am. Ceram. Soc.* **93** 1677–1681
[6] Park N-H, Akamatsu T, Itoh T, et al. 2015 *Materials* **8** (7) 3992–4003
[7] Zavjalov A, Tikhonov S and Kosyanov D 2019 *Materials* **2** (18) 2895–2925
[8] Safronova N A, Kryzhanovska O S, Dobrotvorska M V, et al. 2020 *Ceram. Int.* **46** 6537–6543

AUTHOR INDEX